FIRST MANNED ROCKET LAUNCH

NATTER, THE LUFTWAFFE'S VERTICAL TAKE-OFF INTERCEPTOR

THEN AND NOW

FIRST MANNED ROCKET LAUNCH

NATTER, THE LUFTWAFFE'S VERTICAL TAKE-OFF INTERCEPTOR

THEN AND NOW

by Jean Paul Pallud

AFTER THE BATTLE

An Imprint of Pen & Sword Books Ltd

First Manned Rocket Launch
Natter, the Luftwaffe's vertical take-off Interceptor – Then and Now

Published by After the Battle
An imprint of Pen & Sword Books Ltd
47 Church Street, Barnsley,
South Yorkshire, S70 2AS
Tel. 01226 734222
Fax. 01226 734438
Email: enquiries@pen-and-sword.co.uk
Website: **www.afterthebattle.com**
www.pen-and-sword.co.uk

Printed and bound in India by Parksons Graphics Pvt. Ltd.

ISBN: 9781036101282

Commissioning Editor: Rob Green
Editor: Jean Paul Pallud
Book design: Paul Wilkinson
Jacket design: Jon Wilkinson

Credits:
The major part of this book was originally published as an article in *After the Battle* magazine: 'First Manned Rocket Launch' by Jean Paul Pallud in issue 151 (February 2011).
It has been adapted and enhanced to produce this book and nearly 100 additional photos are included.

Acknowledgements:
The Editor would like to acknowledge the late Manfred Hensel for his invaluable help with the story. The Editor would also like to acknowledge the Bundeswehr and Oberfeldwebel Amelie Baldauf, Hanz Holzer of the Deutsches Museum, Frau Carmen Krug of Hymer AG, Franz Straubinger, Dr Frank Bauer, Thorsten Bretschneider, Sonja Gieschen, Hans Grimm, Fritz Heinzelmann, Hans-Werner Neulen, Christian Payer, Jürgen Sieber, and the late Thomas J. Dietz of the National Air and Space Museum. Finally, the Editor would like to thank Winston Ramsey, who created the wonderful adventure of *After the Battle*, and thus made possible the publication of hundreds of remarkable stories.

Photo Credit Abbreviations:
BA – Bundesarchiv; ECPAD – Médiathèque de la Défense, Fort d'Ivry; USNA – US National Archives. Unless specified otherwise, all illustrations are from the *After the Battle* archive.

Front Cover: *(Top)* In the early days of May 1945, a patrol from Company L of the 324th Regiment reached Sankt Leonhard, in the Pitz valley, and found four Natters lying abandoned by the side of the road. The account that is often seen repeated that the Natters were found at an airfield there has little credence as there is not a single piece of flat ground anywhere in this narrow valley. The white-shirted civilian who willingly demonstrated the technicalities of the Natter to US troops is believed to be Dr. Heinrich Rieck of the DFS. (USNA)

(Bottom) In 2009 the author traced the exact spot where the four Natters were abandoned in 1945 – in a grassy meadow about half a kilometre before reaching the church of Sankt Leonhard.

Back Cover:
Final preparation of prototype M22 before take-off at Truppenübungsplatz Heuberg on February 25, 1945. The test flight was a major step forward in the Natter programme, with the Walter engine and the four boosters all functioning perfectly, as well as the salvage parachutes. (M. Hensel)

CONTENTS

INTRODUCTION

THE NATTER, A rocket-powered vertical take-off interceptor, was born out in the closing months of World War Two, when Germany was on the defensive on all fronts and desperately looking for 'wonder weapons' to stave off the coming of inevitable defeat. This dire situation gave rise to remarkable pioneering achievements such as the first rocket plane, the Messerschmitt Me 163, as well as the first rocket to reach space, the V2, but also to several bizarre devices that gave little or no tactical or strategic advantage. The Natter rocket plane was one of the latter.

Considering the enormous pressure the team was working under, it is quite remarkable that the plane was produced and flown within nine months of the first design work.

It is interesting to speculate on the potential effectiveness of the Natter. Among the questions that arise is whether the pilot would have been temporarily incapacitated by the 2 g acceleration in concert with the noise, vibrations, and the disorienting effect of climbing vertically in the sky. During this initial phase of the flight into the unknown, an autopilot would have flown the aircraft on a predetermined climb path. The pilot would then have taken control to get within range of Allied bombers and have fired 24 unguided Föhn rockets with a single shot. He would have had only a single opportunity to make his attack, there would be no second chance. In response to an interception by a Natter, Allied airstrike planners could have rapidly moved bombers out of harm's way and sent fighter-bombers to destroy the launch site. It is safe to assume that the Natter was a flawed concept from the start and would have been a failure as a bomber interceptor. The fact that none of the countries that had direct access to Natter's plans and technologies in 1945 – the United States, Britain, and France – tried to develop the concept supports this assertion.

Nevertheless, on the cold morning of March 1, 1945, a brave Luftwaffe pilot, Lothar Sieber, volunteered to pilot the first manned launch of a Natter. He became the first man to take off vertically from the ground under rocket power, but the plane crashed and he was killed. The flight of the Natter M23 stands as a major step on the road to spaceflight, even though it has remained virtually unknown to the general public for more than half a century. Long before Yuri Gagarin or Alan Shepard, Lothar Sieber became the first human to take off vertically from the ground under rocket power.

March 1, 1945, prototype M23 of the Natter was ready for launch at Heuberg and Gefreiter Lothar Sieber prepares to swing himself into the cockpit. Minutes after 11 a.m., prototype M23 rocketed up the launch tower and soared skyward until lost to sight in the cloud base. It crashed after 55 seconds of flight, killing Lothar Sieber. (USNA)

FIRST MANNED ROCKET LAUNCH

AFTER OPERATION 'MILLENNIUM', the first 1,000-bomber raid by RAF Bomber Command which targeted Cologne on the night of May 30/31, 1942, the Allied bombing offensive on Germany grew in strength and effectiveness. A five-month-long operation which began in March 1943 against the Ruhr industrial area was followed at the end of July by Operation 'Gomorrah', a major offensive against Hamburg, Germany's second largest city. Hamburg was attacked for eight days and seven nights, huge firestorms killing 40,000-50,000 civilians. Over 60 per cent of Hamburg's living accommodation was destroyed or made uninhabitable and in the immediate wake over two-thirds of the population fled the city. These attacks by both the RAF and USAAF struck a major blow at German morale and severely shook the Nazi leadership, leading Hitler to be concerned that similar 'terror' raids on other cities could force Germany out of the war.

Although Luftwaffe fighter squadrons were doing their best to inflict casualties, and on some raids caused serious losses, they were still unable to counter the force of the Allied offensive. A radical solution was desperately needed.

Back in September 1942, the Luftwaffe Flak commander, Generalmajor Walther von Axthelm, had already asked for anti-aircraft missiles and his request had been developed by several manufacturers for some months. Both the Henschel Hs 117 'Schmetterling' and the Rheinmetall-Borsig 'Rheintochter' were test-flown in 1943, and the EMW 'Wasserfall' and Messerschmitt 'Enzian' systems were also soon to be tested. However, although the concept appeared promising, it was clear that a long development period would still be required before the problems of missile guidance and fusing were solved. (No efficient guidance system of the anti-aircraft missiles reached service application before the end of the Second World War).

The idea of providing the missile with a pilot who could control the weapon during the final phase of the attack might be a short-term solution and, though the steps that then occurred within the Reichsluftfahrtministerium (German Air Ministry or RLM) are not precisely known, in the spring of 1944 the ministry issued requirements for a rocket-propelled, point-defence interceptor.

Four aircraft manufacturers, Messerschmitt, Junkers, Heinkel and Bachem,

A graduate from the Technischen Hochschule of Stuttgart, Dipl.Ing. Erich Bachem designed gliders and caravans (the latter he also resumed after the war) and from 1933 he was the Technical Director at the Gerhard-Fieseler-Werke, and from 1938 Chief of the Development Department. Some of his developments for Fieseler were patented. This photo of him with Hanna Reitsch, the famous aviator and test pilot, was taken during a flying day in Kassel-Waldau on July 17, 1938. (M. Hensel)

produced designs, all to be propelled by the same engine: the Walter 109-509 liquid propellant rocket unit. Three of the competitors, the Messerschmitt Me P 1104, the Junkers EF 127 'Walli' and Heinkel P 1077 'Julia', were somewhat conventional as they were designed to land back on skids.

The fourth proposal by Dipl.Ing. Erich Bachem was much more radical. His 'Natter' was to be a semi-expendable vertical take-off interceptor

Bachem-Werk engineers at work in 1944. From L-R: Fahrnbauer, Erich Bachem, Gebhardt Zorell, and Wolfgang Offik. Offik was the manager of the day-to-day development and construction matters relating to the Natter project.
(M. Hensel)

launched from a ramp. Once it reached the bomber stream, the pilot would fire a salvo of rockets at one bomber before gliding away. The machine would then separate in mid-air, the pilot and the rear part with the rocket engine landing individually by parachute. Another feature of the Natter was its semi-expendability, low-grade, non-essential materials being used throughout, and assembled by semi-skilled labour.

The idea was not entirely original in that Dr. Wernher von Braun had already proposed the concept of a rocket-powered, vertical take-off interceptor in 1939 but the RLM had rejected it at the time as they considered it offered no tactical advantage. Von Braun re-submitted the concept in May 1941 but the RLM again rejected it.

Also, the suggestion of a piloted 'ram rocket' was mooted in September 1943 by Dr Alexander Lippish. His proposal was for a rocket-propelled aircraft that would take off vertically and quickly climb to the altitude of the bombers, reaching 10,000 metres in 40 seconds. The pilot would steer the interceptor to impact with a bomber, its solid, sharp ramming point slicing through the airframe, breaking it apart. With its speed and trajectory significantly unchanged by the impact, the rocket would continue to climb away, finally decelerating to zero velocity at the top of its trajectory. The pilot would then release a parachute which would recover the interceptor while the pilot would bale out separately at a lower altitude.

To follow up his radical proposal, Bachem visited a number of

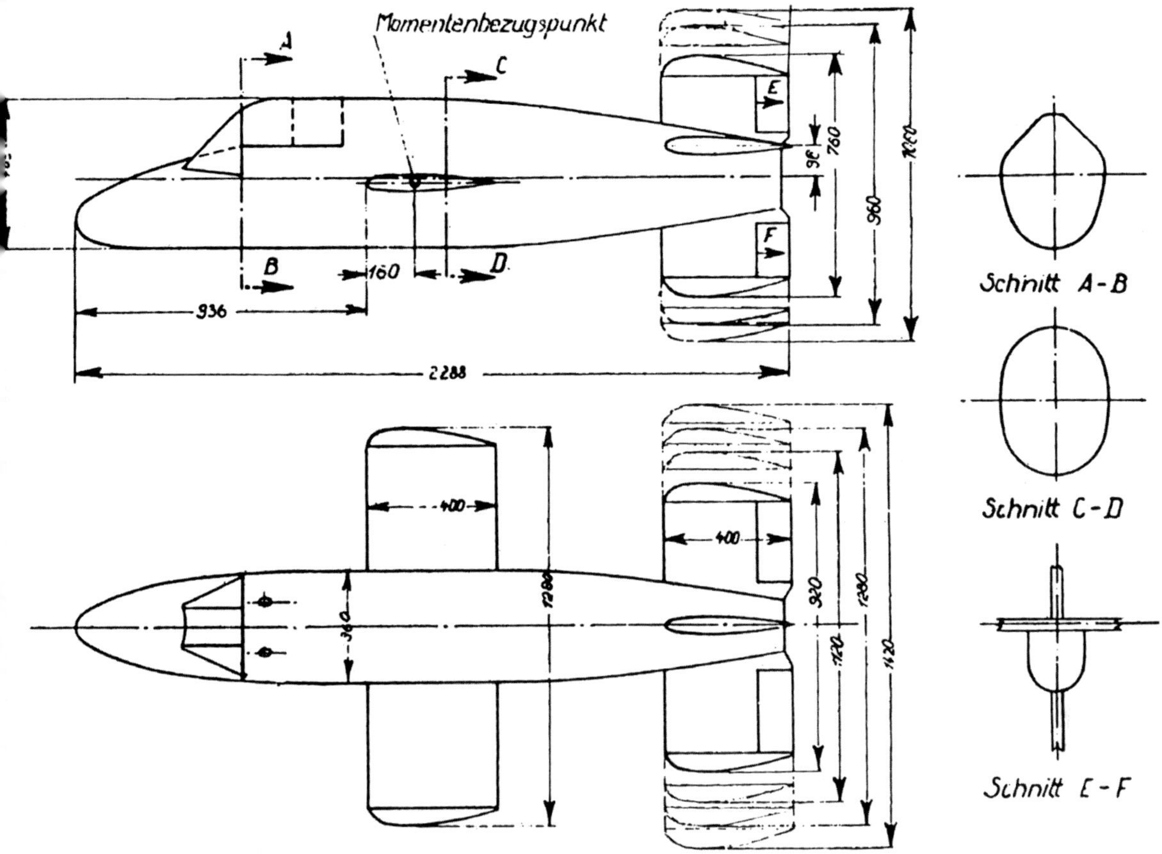

Abb. 1
Übersicht für BP 20–07

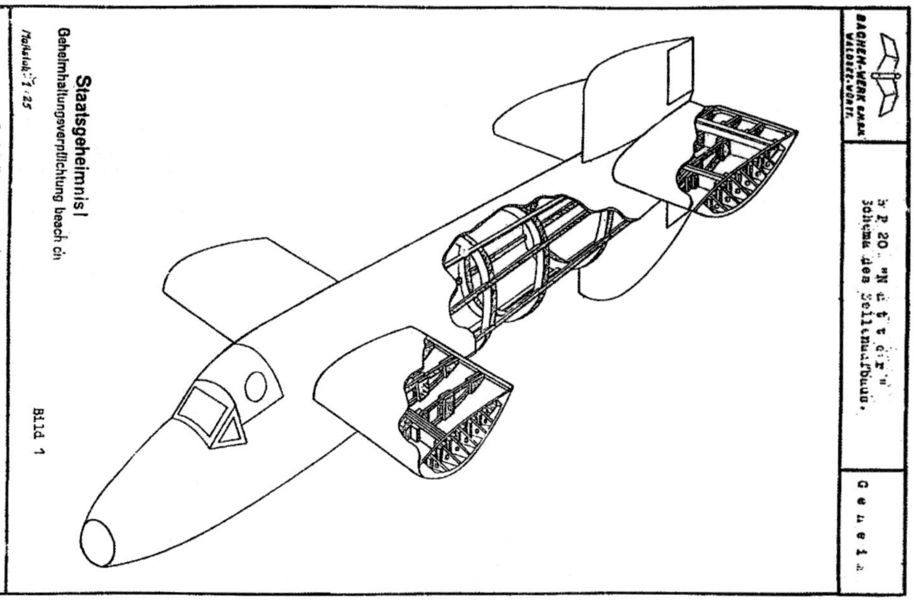

The Natter measured 6.06 metres long and had a wingspan of 3.65 metres, the take-off weight being 2,200 kilos (this data referring to the A1 model). The wings were plain rectangular wooden slabs without ailerons or flaps. The cruciform tail had four adjustable fins, the various combinations of which controlled pitch, yaw and roll once the Natter had reached sufficient speed to generate adequate airflow.

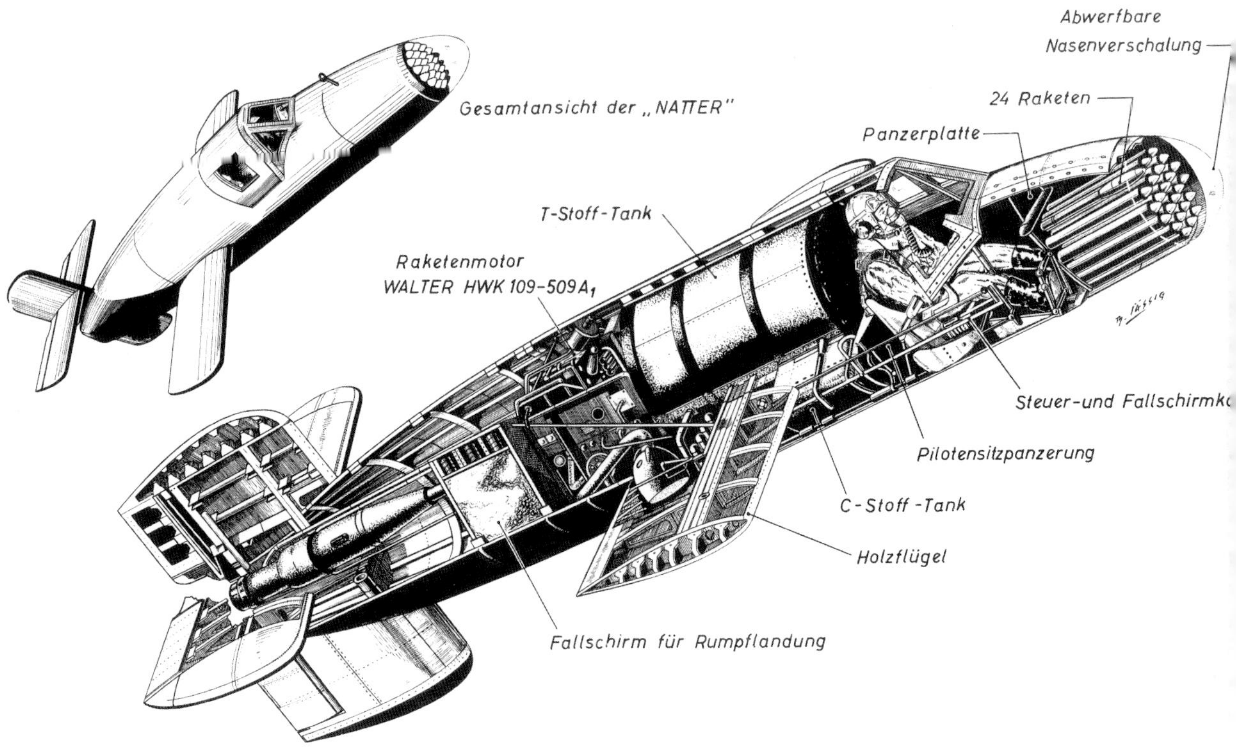

The armament in the nose comprised a battery of unguided rockets (the alternative armament of 24 R4M rockets is shown here). The two fuel tanks were aft of the rear cockpit bulkhead, one above the wing spar accommodating 365 litres of T-Stoff, the other below containing 165 litres of C-Stoff. The Walter HWK 109-509 main engine generated a maximum thrust of 1,700 kilos but it could be throttled back to 300 kilos. Bolted to the rear fuselage were four jettisonable Schmidding 109-533 rocket boosters, two per side, which provided a combined thrust of 4,800 kilos for ten seconds. At lift-off, with all five motors ignited, the thrust generated was 6,500 kilos and the resulting acceleration about 2g. After ten seconds of vertical flight, the exhausted boosters were released automatically and fell away. At that point the interceptor had reached an altitude of 1,200 metres and a speed of 880 km/h. The auto-pilot then engaged to alter the flight path to an angle of 60 degrees above the horizontal and the machine rocketed to an altitude of 12,000 metres in a remarkable 60 seconds. Combat time at this altitude with the motor throttled back to 300 kilos, would have been less than two minutes. (This post-war sketch was drawn by Theodor Lässig. Note the boosters are not shown.)

influential people in the summer of 1944 to enlist their support, including Generalleutnant Adolf Galland, the General of the Fighter Arm, who had shown interest. However, as the tendering ended in August, there had not been sufficient time for Bachem and his team to produce detailed drawings of their project. The plans they submitted were somewhat primitive and, not surprisingly, were rejected by the RLM. Instead, they opted for the more conventional proposal from Heinkel – the Heinkel P 1077 'Julia' – that would take off from an inclined rail and land on a skid.

Nevertheless, Bachem refused to give up and sought the support of Reichsführer-SS Heinrich Himmler, apparently via SS-Obergruppenführer Hans Jüttner whom Himmler had appointed Chef der Heeresrüstung und Befehlshaber des Ersatzheeres (Chief of Army Armament and Commander of the Reserve Army) in July 1944 after he replaced Generaloberst Friedrich Fromm following the attempt on Hitler's life. Bachem was invited to Berlin

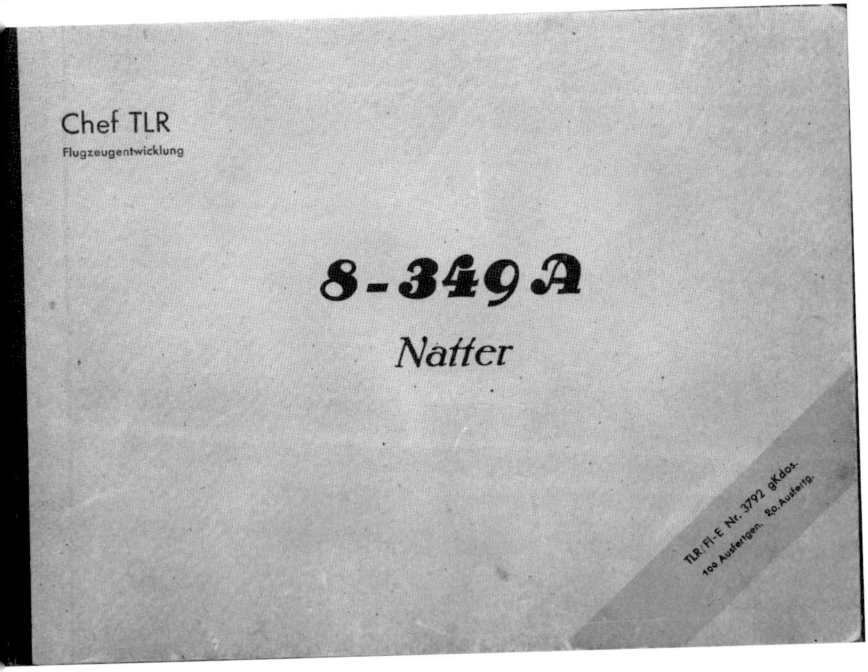

At the end of October 1944 RLM gave the Natter project the number 8-349. These sketches are taken from a report by the head of the Luftwaffe Technischen Luftrüstung. This copy now in the Heuberg military museum at Stetten am kalten Markt shows how the Allied technical investigation team pasted English translation directly onto the pages of the original German document.

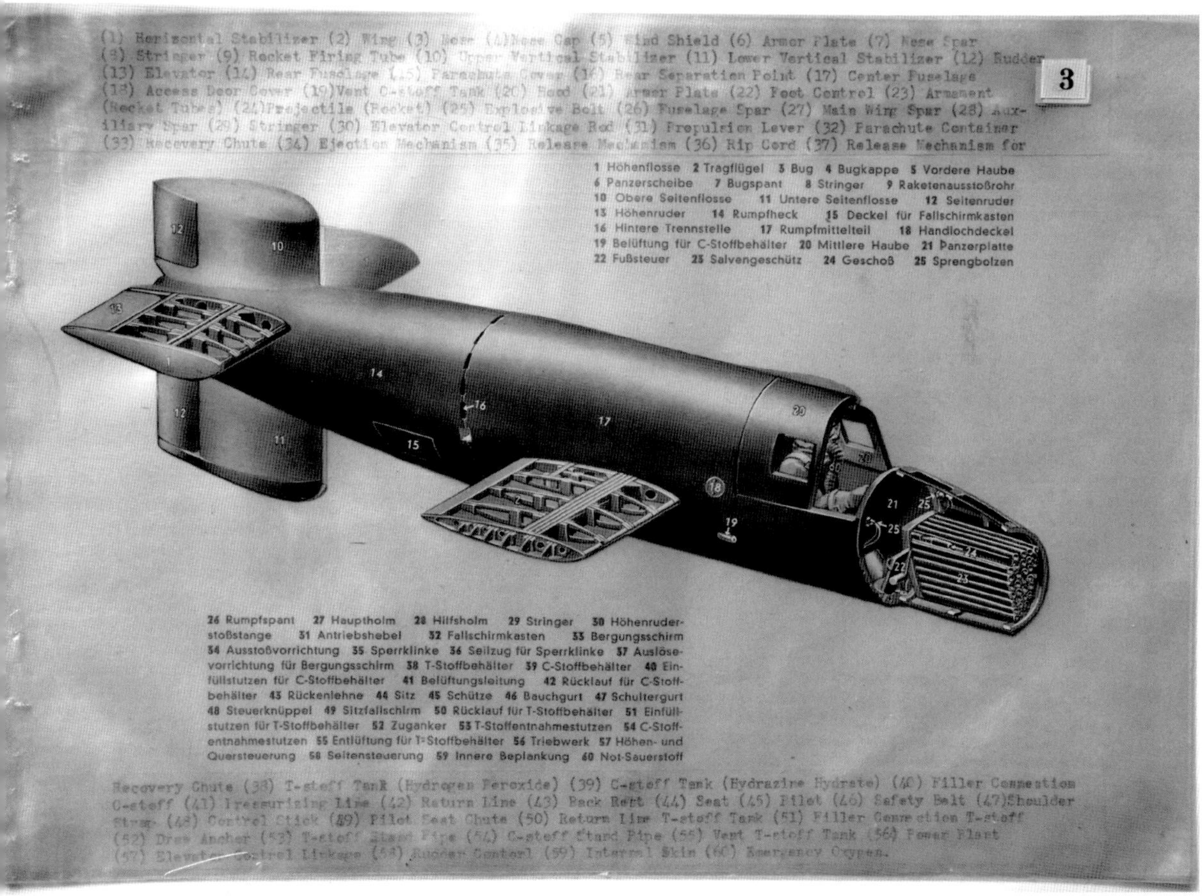

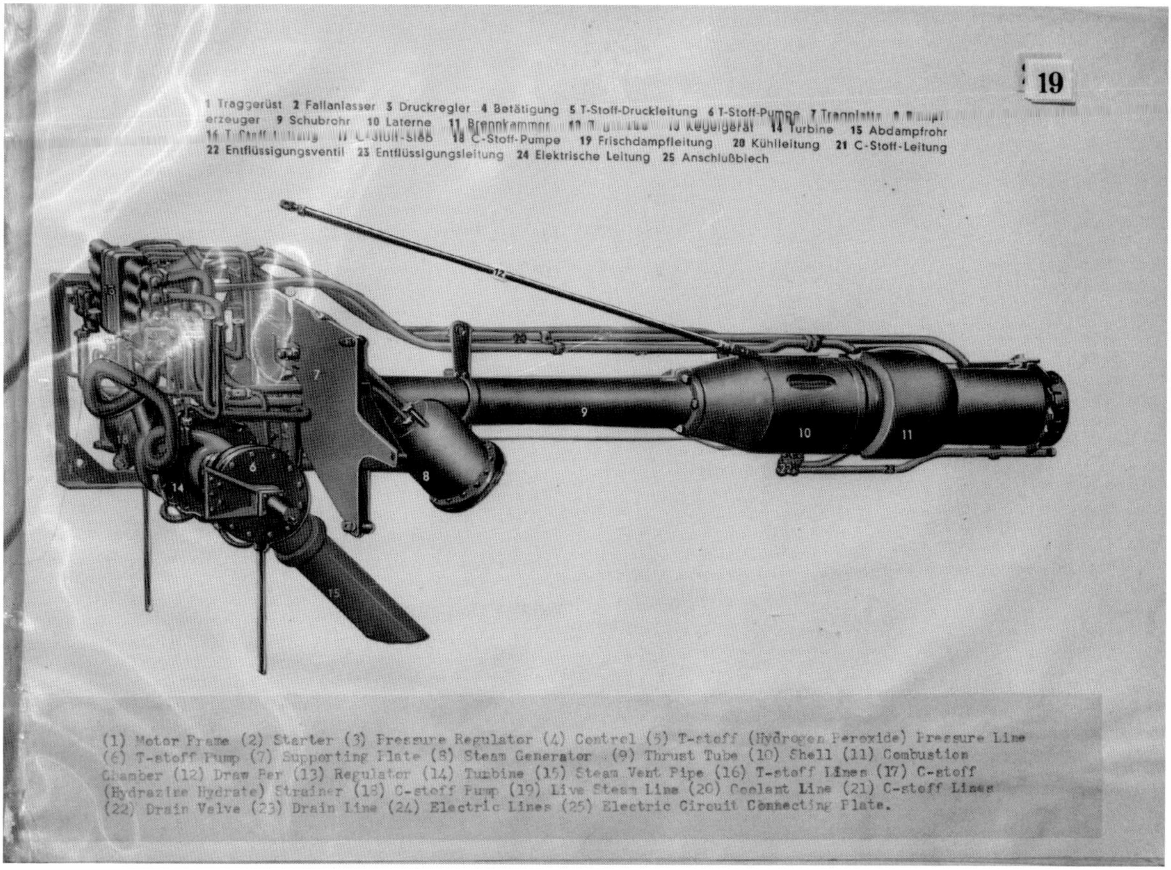

From this same document, the description of the Walter 109-509A rocket motor. Two liquid fuels – T-Stoff (a highly caustic solution of hydrogen peroxide and a stabilizing chemical) and C-Stoff (a hydrazine hydrate/methanol/water mixture) – combined inside the motor to generate thrust. As combustion was spontaneous when T-Stoff mixed with C-Stoff, extreme care was required in handling both chemicals.

for an interview with Himmler and by mid-September the Technical Office of the Waffen-SS issued Bachem with an order to develop and build his one-shot interceptor. (A post-war report stated that by 1945, 150 rockets had been ordered by the SS and 50 by the Luftwaffe.)

The name 'Natter' was an early designation by Erich Bachem but the factory designation for the project was BP-20, possibly an abbreviation for Bachem-Werk Projekt No. 20. The RLM had no alternative but to support the project and by the end of October had given it the official type number 8-349.

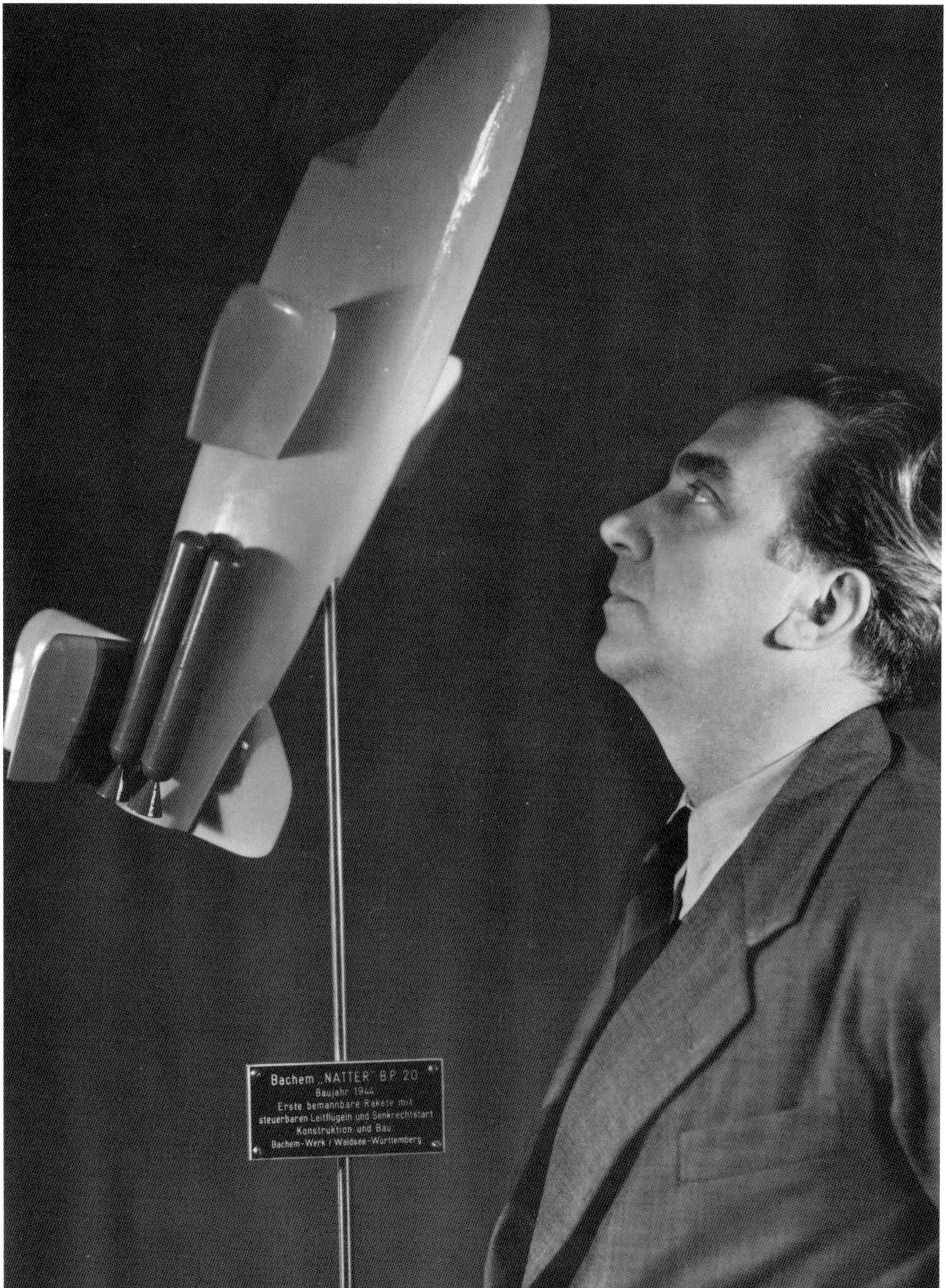

A post-war photo of Erich Bachem looking at a model of the Natter.

MISSION PROFILE

THE ROCKET INTERCEPTORS were to be based in the vicinity of the installations that they were to protect. Each aircraft was to be ready on the launch ramp, loaded with its propellants and, when the approach of a bomber formation was picked up by radar, a warning was sounded for the pilot to get into the cockpit. The ground crew rotated the ramp and clamped it at the angle directed by the flight control station.

The radical mission profile of the Natter concept as described in the report compiled in July 1945 by the US Naval Technical Mission in Europe under Dr. Clark B. Millikan: 'The plane is attached to the launching tube

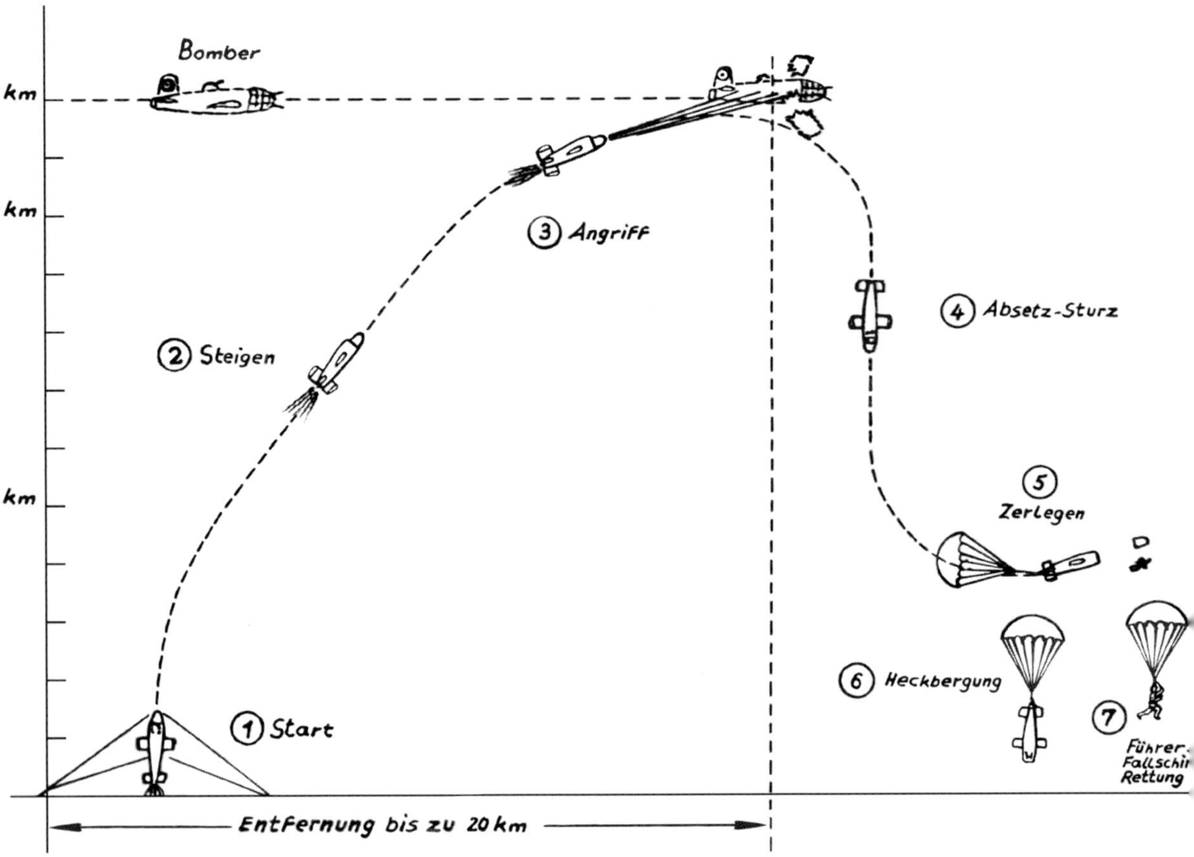

on which it can slide vertically and around which it can rotate. A standard anti-aircraft director is placed alongside and the orientation of the plane is set as is the elevator deflection required at the end of the vertically flight so that a straight flight path will reach the predicted location of the enemy bomber.

The motor and launching rockets are fired and the plane ascends vertically some 150 metres when the auto-pilot deflects the elevators and holds the plane on the predetermined straight, climbing flight path. At this time the boost rockets are dropped. When the pilot sights the enemy, he takes over and when on the target and in range fires the 24 rockets. He then dives away and when clear slows to about 250 km/hr and releases the nose section which is sucked forward and clear by the pressure distribution over it. The pilot then releases a parachute from the rear of the fuselage. The latter decelerates very rapidly ejecting the pilot who parachutes to the ground.'

---

[1] Once the bombers were within range, the command 'Fliegeralarm' was given. The pilot clipped on his oxygen mask, switched on the on-board power supply and the auto-pilot. He started the turbine that reached its nominal speed after a few seconds when the Walter rocket engine then roared into life. He pushed the throttle to full, the turbine reached 16,000 rpm and the rocket engine developed its maximum thrust of 1,700 kilos. The pilot firmly grasped two handgrips and with his left thumb he pressed the button triggering the ignition of the four Schmidding booster rockets.

[2] The interceptor rose sharply to the vertical with an acceleration of 2 g. After ten seconds of climbing straight up, the exhausted boosters were released automatically and fell away. The interceptor was by then at an altitude of 1,200 metres, the pilot still being a passive passenger. At this moment the auto-pilot engaged to incline the flight path to an angle of ascent 60 degrees from the horizontal. The progressive reduction in the weight of the machine with the depletion of the propellants would result in a gradual increase in speed so the pilot manually throttled back the motor to maintain a constant velocity of about 800 km/h.

[3] Within about a minute the interceptor would have reached the altitude of the bomber formation, about 10,000 metres. The fighter was now 1,500 metres from the bombers and below, rapidly closing on them. The pilot then turned off the auto-pilot and took command. He manoeuvred to bring one bomber into his gun-sight and kept on until its wingspan reached the pre-determined position in his sight. The aircraft would then be about 300 metres from its target and the pilot pressed the firing button. Explosive bolts holding the nose cap detonated; it fell away and the 24 rockets were all fired in one half-second salvo. The Natter flew straight on across the bomber formation.

[4] Its fuel exhausted, the Walter motor cut out. The pilot pushed the control stick and dived down away from the combat zone. He pulled out of the dive at an altitude of some 3,000 metres and slowed to about 250 km/h over the pre-planned recovery area. He then prepared to leave the interceptor. He released the canopy, disconnected his intercom, undid his seat belt, and unlocked the control column so that he could tilt it forward. He then operated the lever that released the nose section from the body of the interceptor.

[5], [6] and [7] The nose flew off with the windshield, instrument panel and rudder pedals and, as it did so, the nose pulled two cables that opened the parachute hatches and triggered the spring-loaded ejection mechanism. The salvage parachute opened, suddenly decelerating the interceptor and, under the shock the pilot was thrown out and fell freely. He opened his parachute and floated down.

DEVELOPMENT

DESIGNS WERE SPEEDILY worked on throughout September and October at the Bachem factory, the head of the Natter project being Dipl.Ing. Henri Frans Bethbeder, a Dutchman who had been a Technical Director at Dornier before joining Bachem. Several different designs were worked upon simultaneously to cover the various concepts. The prone position for the pilot was abandoned at the beginning of October, opting instead for a conventional seated position. However, Bachem kept true to the basic requirements of simplicity using glued and screwed wooden parts. He later claimed that the construction cost of his interceptor would amount to about one fiftieth of that for a normal fighter aircraft, including the propulsion unit and equipment.

A timetable was prepared – decidedly optimistic – with the immediate

Aerial view of the Bachem-Werk during the war, with the main production hall at top centre and gatehouse centre left. Biberacher Strasse can be seen top left. (H. Grimm)

Bachem left Fieseler in 1942 to establish his own company at Waldsee, 40 kilometres north of Lake Constance, to provide spare parts for the aircraft industry. Partner and co-founder was Willy Fiedler, a friend from their university days. The company was established on February 10 and took over a large production hall left over by a sawmill which closed in the 1930s. The premises had their own branch line to the rail network. (H. Grimm)

The entrance of the Bachem-Werk factory as seen during the war with the gatehouse in the centre and the main production hall in background. Located at No. 98 Biberacher Strasse, Waldsee, the signpost read 'Bachem-Werk, Wood and Metal Construction'. The company name 'ERIBA' was taken from Erich Bachem's nickname in his student days, namely ERIch BAchem. (H. Grimm)

The old main production was still there Biberacher Strasse when the author visited Waldsee in 2010 and was then used for storage by HYMER, the well-known caravan and motorhome manufacturer. (ATB)

construction of 20 experimental machines, soon extended to 50. Pilot production of 200 interceptors was to start in November and full production was planned to begin in February 1945. A monthly output of 3,000 to 5,000 machines was envisaged.

Himmler quickly mobilised a workforce and teams of skilled workers soon joined Bachem-Werk at Waldsee. Known as the Waffen-SS Sonderkommando Waldsee, one company of about 120 technicians under the command of SS-Obersturmführer Heinz Flessner was soon hard at work. In addition, the factory employees comprised a mixture of Germans,

Men of Waffen-SS Sonderkommando Waldsee on parade in front of the production hall (out of the picture to the right). The entrance from Biberacher Strasse is in the background. Under the command of SS-Obersturmführer Heinz Flessner, this special detachment of Waffen-SS comprising unarmed technicians and skilled workers, had a strength of about 120 men at Waldsee. (H. Grimm)

mainly women, prisoners of war and deportees. The latter were housed in huts in the factory grounds. By the end of 1944, Bachem-Werk had a total of around 600 workers on site.

In September, tests were made with a 1:2.5 scale wooden model of the BP-20 airframe in the wind tunnel of the Deutsche Versuchsanstalt für Luftfahrt (DVL) at Berlin-Aldershof airfield. Further tests were carried out using the high-speed wind tunnel of the Luftfahrt-Forschungs-Anstalt Hermann Göring (LFA) at Braunschweig which could produce speeds

Factory staff working on the construction of the tail units. (M. Hensel)

General view at the Bachem-Werk showing fuselages being covered with 3mm beech plywood glued and nailed to the main framework. (H. Grimm)

Another view of the production hall, this time with A1 production Natters under construction in March 1945. (M. Hensel)

When the author visited Waldsee in 2010 he learnt that the building was scheduled for demolition but Hymer AG kindly agreed to have photos taken for us in the historical Natter construction hall before it was demolished.
(M. Hensel and Frans Straubinger)

close to Mach 1. The tests proved that the Bachem interceptor would be satisfactory up to speeds of 1,100 km/h.

The first prototype of the BP-20 was completed on October 4, and the operational design of the interceptor was first defined about the same time as a blueprint dated October 3 has survived. Several variants would evolve over the following two months, named BP-20 A, B and C.

Bachem-Werk, Then and Now. Rear sections of three of this same series of the A1 operational machines, two numbered 8 and 9, and a nose section in the foreground.
(H. Grimm and Frans Straubinger)

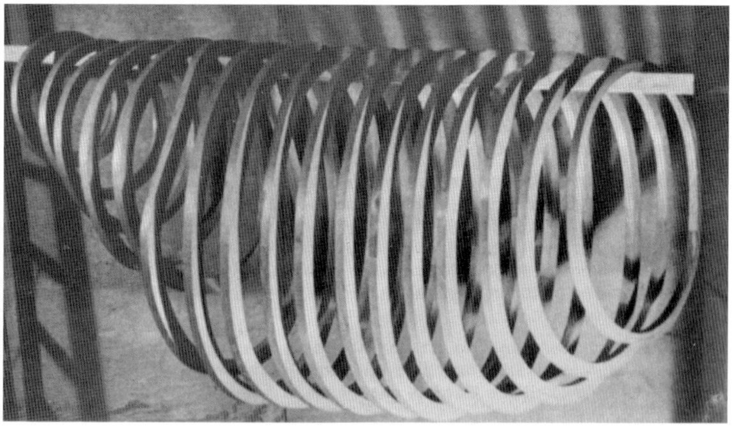

This photo shows 17 of the fuselage hoop frames ready for assembly: the fuselage contained 19 frames, five in the nose, eight in the mid-section and six in the rear fuselage.
(M. Hensel)

A series of views showing the work in the construction hall of the Natter: fuselages under construction in a jig.
(M. Hensel)

Wings under construction, note the main spar composed of four wooden girders. (M. Hensel)

A photograph taken looking backwards into a partially completed fuselage, showing the control column and, in foreground, the rudder control linkage. (M. Hensel)

Another shot looking aft into the fuselage, showing the push rod system operating the elevons in the tail unit. (M. Hensel)

The 60 mm thick armoured windscreen frame was bolted to the fuselage. The plexiglass side windows were 10 mm thick. (M. Hensel)

Initially, the pairs of jettisonable Schmidding 109-533 rocket boosters bolted to the rear fuselage were spaced apart by chocks and strapped together. (M. Hensel)

Subsequently, the booster rockets were mounted separately to the side of to the rear fuselage. (M. Hensel)

Close-up of the honeycomb canister for 24 Föhn rockets in the nose section of the Natter. In operation, all the 24 missiles were fired in ripple over a period of 0.3 seconds and on each side of the plane one panel was blown away to allow the exhaust gases to escape. (M. Hensel)

With a length of 28 cm and weighing 2.74 kg, the Föhn was spin-stabilized. The rocket motor consisted of a single stick of solid-fuel propellant and the exhaust gasses were forced through seven straight inner venturis and seven angled outer venturis which imparted spin. The warhead contained 280g of explosive, with a nose-mounted impact fuse exploding on contact and a time-delayed fuse that was initiated during ignition of the rocket motor.

Bachem-Werk personnel at an airfield in the Schussen valley near Baienfurt in October 1944 during the testing of the 'Lerche' motorised ultra-light aircraft. From L-R: Eugen Walter, Friedhelm Christ, Gebhardt Zorell, Henri Frans Bethbeder, Paul Wilbrett (standing), Fahrnbauer, Erich Bachem, Hans Steybe, Willy Fiedler and Fritz Blessing. Bethbeder was a Dutchman who had also studied in Stuttgart and was now head of the Natter project at the Bachem-Werk. Fielder, who from September 1942 was the leader of the industrial testing of the V1, was to join the Bachem-Werk full time early in 1945 to assume the position of Chief Designer. In 1948, he moved to the US to work at the Naval Missile Test Center in California, and in 1956 joined Lockheed Aircraft where he worked on the US Navy's Polaris ballistic missile. (M. Hensel)

The initial ten prototypes were devised to test the flying qualities of the machine, beginning with manned glider tests. The prototype BP-20 M1 was taken to Neuburg an der Donau, about 70 kilometres north of Munich, where flight-testing would be carried out under the auspices of the Deutsche Forschungsanstalt für Segelflug (DFS). The test machine did not include an engine and was to be towed up to 3,000 metres by a Heinkel He 111. Not having an undercarriage, it had to be mounted for take-off on a tricycle trolley.

The towed test flight of prototype M1 was carried out on November 3 with Erich Klöckner of the DFS at the controls. Although the flight went well, it appeared that the aircraft was somewhat unstable on its transverse axis. Klöckner bailed out as planned before the Heinkel attempted to land with the unmanned interceptor but it hit the ground and was destroyed.

Prototype M3 (fitted with a tricycle undercarriage) was completed on November 20 and it made its maiden flight on December 14, again with

To assess the flight characteristics of the Natter, manned glider tests were carried out at Neuburg-an-der-Donau, about 70 kilometres north of Munich, under the auspices of the DFS (Deutsche Forschungsanstalt für Segeflug). The BP-20 M1 prototype was mounted for take-off on a tricycle trolley. (M. Hensel)

The M1 prototype was prepared for flight on November 3, 1944. From left to right: Dipl.-Ing. Willy Fiedler, Dipl.-Ing. Zacher, DFS head of the Natter glider testing and test pilot Erich Klöckner. Note the Messerschmitt Me 262 in the background. (M. Hensel)

Piloted by test pilot Klöckner, the Natter was to be towed up to 3,000 metres by the Heinkel He 111 tow plane seen here in left background. (M. Hensel)

The M1 ready for take-off with the tow cable attached near the wingtips (about 60 metres long, the cable was bifurcated about six metres in front of the Natter to be attached to each wingtip). The test flight went well and Klöckner bailed out as planned. The Heinkel then brought back the unmanned M1, still in tow, to try to land it but the machine was smashed during several wing-overs after being released over the airfield. (M. Hensel)

Fitted with a tricycle undercarriage prototype M3 was completed two weeks after the flight of prototype M1. (M. Hensel)

M3 made its maiden flight on December 14 with Erich Klöckner in the cockpit. The suspension points of the towing harness, which had been traced as the cause of the M1 instability during landing, had been adjusted and the flight was successful. Klöckner bailed out and the Heinkel gently landed the unmanned M3. (M. Hensel)

Erich Klöckner in the cockpit. The suspension points of the towing harness, which had been traced as the cause of the M1 instability, had been adjusted and the flight was successful. Klöckner bailed out and the Heinkel landed the unmanned M3 without damage. Another test flight is believed to have taken place on December 22 (although no report on it appears to have survived) and another on January 27, 1945.

The results being positive, it was decided that time had come to conduct a manned free flight. Prototype M8 was made available with a Luftwaffe test

In December, tests were carried out with the M5, M6 and M7 prototypes to check the nose separation system and the parachute opening. The Natter was slung under the right wing of a Heinkel and dropped from an altitude of 500 metres over the Truppenübungsplatz Heuberg (military training area Heuberg) near Stetten am kalten Markt. A dummy pilot was fitted with parachute and seated in the cockpit. (M. Hensel)

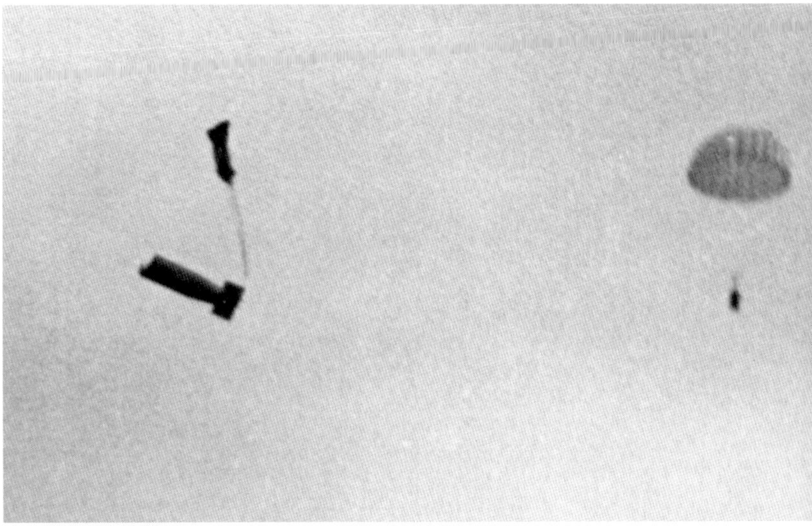

One of the unmanned Natter prototypes – possibly M5 or M7 – after having been dropped over the Heuberg training ground to test the nose separation system and parachute deployment. The tests generally went well, the dummy pilot coming safely to earth on its own parachute. (M. Hensel)

Damage to the main fuselage that landed under their parachutes was light. This is apparently M5. (M. Hensel)

pilot, Unteroffizier Hans Zübert, at the controls. On February 14 a Heinkel towed him to an altitude of 5,500 metres at which point Zübert released the coupling to begin the test programme. He dived to a speed of 600 km/h, tested the stalling speed around 200 km/h, and executed turns both up and down and side to side. Then, at an altitude about 1,200 metres, he baled out. His report on the flying characteristics of the BP-20 were quite favourable.

On February 14, a Heinkel He 111 towed the M8 prototype to an altitude of 5,500 metres with test pilot Hans Zübert in the cockpit. He tested the flight characteristics of the machine in a free flight and found them to be quite good. At the end of the flight program, Zübert attempted to activate the nose separation and parachute system but it failed to operate. He put the Natter into a steep climb to reduce speed, released the cockpit enclosure and bailed out in the normal manner. He opened his parachute and landed safely. The Natter crashed to the ground. In his report of July 1945 Dr. Clark B. Millikan noted as follows the flight characteristics as reported by Zübert:

(1) Stability was excellent and controls light and well coordinated for indicated air speeds between 200 and 700 km/hr.
(2) There was no rolling moment due to sideslip, and no apparent yawing moment due to differential deflection of the elevators to produce roll.
(3) The rate of roll was estimated at 1 rev. per second.
(4) At 400 km/hr a full circle could be turned in approximately 20 seconds.
(5) The controlled stalling speed was 200 km/hr indicated air speed which occurred at an angle of attack of about 30°.
(6) The handling and flying qualities were judged by the pilot to be superior to those of any of the standard German single-seat fighters.

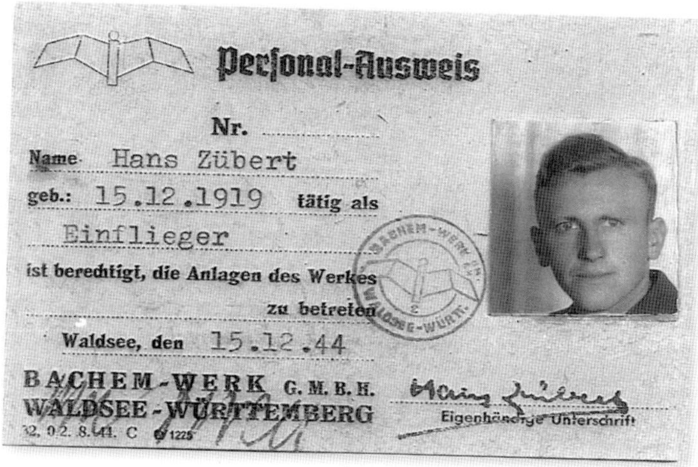

This is Hans Zübert's identity card at Bachem-Werk. (M. Hensel)

The Föhn armament during a test: the 24 Föhn missiles were ripple fired over a period of 0.3 seconds, and a panel (the smaller one seen here) was blown away on each side to allow the exhaust gases to escape. The larger opening was to access electrical connections for the test. (M. Hensel)

The A1 series of operational Natters produced from March 1945 were equipped with the Föhn rockets canister in the nose but an alternative armament system was also tested at the Heuberg's firing range. The Rohrbatterie 108 consisted of a cluster of 32 cannon barrels each containing a 30 mm shells. The shells were fired in a sequence over a period of 0.2 seconds. (M. Hensel)

However, one problem remained: what would be the change in the position of centre of gravity of the machine during the flight as sudden shifts would occur when the boosters were jettisoned and the weapon rockets were fired?

In the meantime, a separate series of tests were being conducted to check the process of releasing the pilot at the end of the mission – the one really novel aspect of the Natter concept. To test the nose separation system and parachute opening, prototypes were to be slung under the right wing of a Heinkel and dropped from an altitude of 500 metres. Detonation of explosive bolts holding the nose section to the fuselage would initiate the nose separation. A dummy pilot equipped with parachute was seated in the cockpit.

Prototype M2 was thus tested at DFS Ainring airfield near Salzburg in November but the bolts failed to fire and M2 crashed to the ground. A further series of three tests were then carried out in December with prototypes M5, M6 and M7 being dropped over the Truppenübungsplatz Heuberg (Military Training Area Heuberg) near Stetten am kalten Markt, 50 kilometres north of Lake Constance. These tests generally went well, the dummy pilot coming safely to earth on its own parachute, which confirmed the principle of releasing the pilot through the action of the separation of the nose section.

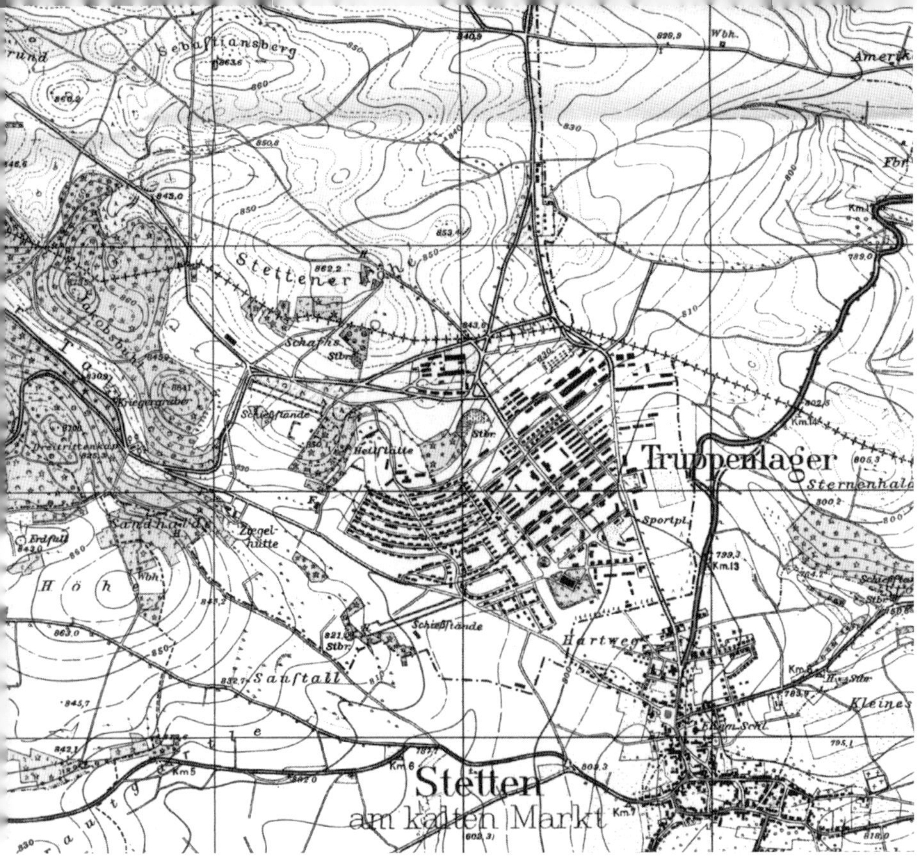

The Truppenübungsplatz Heuberg training ground at Stetten am kalten Markt was first established in 1910. During the First World War it was also used as a POW camp and by 1917 there were 5,000 soldiers undergoing training there alongside 15,000 prisoners of war. After the war part became a children's home and hospital, and for nine months in 1933 it was a concentration camp before being taken over the following year by the Army. During the Second World War it was used as the mobilisation camp for army units like the 4. Gebirgsjäger-Division in late 1940 and the Afrika-Brigade 999 in 1942.

In the summer of 1944, four Italian divisions of the Italian Social Republic, Mussolini's puppet state, were assembled in Germany. Some elements were trained in Heuberg where Mussolini came in person to encourage them on July 17, 1944. Later in 1945 the Indische Freiwilligen-Legion and the Russian 650. Infanterie-Division were organised and trained at Heuberg. French troops reached the Heuberg camp on April 22, 1945, freeing 20,000 Russian prisoners of war. The French army took over the Heuberg camp in 1945, joined from 1957 by units of the new Bundeswehr. In the early 1960s, elements of the US 357th Artillery Detachment are said to have kept nuclear warheads at the Heuberg camp which would have been issued to French Nike Hercules missile units had a war with the Soviet Union begun. The French presence ended in 1997 when the 3ème Régiment de Dragons departed. (M. Hensel)

In 1980 Hauptfeldwebel Manfred Hensel, the telecommunication officer at Heuberg, published Chronik des Truppenübungsplatzes und der Garnison Heuberg, a thoroughly researched history of the camp and its garrisons. (M. Hensel)

In the meantime, a launch site to undertake powered vertical take-off trials had been established at the Heuberg camp on an open piece of ground called the Ochsenkopf. The main structure was a 20-metre-high launch tower with three guide rails for the wingtips and lower fin to stabilise the Natter until it cleared the tower. The whole launch process was spartan as

To undertake vertical take-off trials of the Natter, a 20-metres-high launch tower was built on the Ochsenkopf, a high open field at the Heuberg training camp. A first launch was attempted on December 18 with the M21 prototype. It was not fitted with the Walter engine and was therefore only to be launched by the four booster rockets. Bachem (centre) and Dr.Ing. K. Pertikat mixed with the ground crew at the bottom of the launch tower as M21 was prepared for launch. (M. Hensel)

Wolfgang Offik, Erich Bachem and Hans Steybe. (M. Hensel)

Camera and binoculars at the ready, Bachem took one last look at M21 before launch. The four Schmidding rockets all ignited but the launch latch failed to release the aircraft as expected and the M21 eventually caught fire and burned. (M. Hensel)

The fault with M21 was attributed to electrical cables having been burned by the rocket exhaust. They have been repositioned so that they were safe from the exhaust and a mechanical back-up release mechanism was installed. Not equipped with the Walter motor, and hence to be launched only by the four booster rockets, prototype M16 was prepared for launch. (M. Hensel)

the delicate process of filling the liquid propellants was achieved using a ladder resting against the wing.

A first launch was attempted on December 18 with prototype M21, albeit without the Walter motor. The four booster rockets ignited according to plan but the latch to free the interceptor failed and the Natter remained firmly stuck at the bottom of the launch tower. The tail soon caught fire and the whole machine was destroyed.

A fault in the ground equipment was traced and cured and a new test set up on December 22 with prototype M16. Although one of the booster rockets failed to ignite correctly, the Natter flew upwards in a stable parabolic trajectory culminating at 800 metres before crashing about 1.5 kilometres from the launch site.

Prototype M17 was launched on December 29, on this occasion fitted with a time-delay mechanism to release the salvage parachute. All four booster rockets ignited perfectly and the machine reached an altitude of

This photo of M16, that is originally in colour, shows that it was painted in RLM 05 ivory, a finish typically used on wooden aircraft like gliders, with the rocket boosters in RLM 23 red and the bands at the separation points RLM 22 black. (USNA)

Engineers are attending to the tethering rod that was designed to break away when the thrust reached 1,800 kilos and ground crew lay electrical cables. (M. Hensel)

The M16 prototype was successfully launched on December 22. Although one of the booster rockets failed to ignite properly, the Natter flew upwards in a stable parabolic trajectory peaking at 800 metres before crashing about 1.5 kilometres from the launch site. (M. Hensel)

In another test with only the four Schmidding booster rockets, the M17 prototype was readied for launch on December 29. It was painted in RLM 04 Yellow with a sunburst pattern on the wings in RLM 22 Black. This was to allow an easier analysis of the movements of the aircraft on the photos and films taken during the flight. (M. Hensel)

The booster rockets of M17 burnt perfectly, though later analysis of the film taken during the test showed that their thrust was below the rated 1,200 kilograms. The machine reached an altitude of 2,500 metres and the time delay mechanism released the recovery parachute as planned but the cable tore off at an altitude of 700 metres and the machine crashed. (USNA)

some 2,500 metres. Although the parachute was released as planned, it ripped off causing the Natter to crash. Analysis of film taken during the test confirmed what had been revealed during the flight of M16: that the thrust of the booster rocket was only about 1,000 kilos, well below the rated 1,200 kilogrammes.

Whatever the reason (possibly because of the bleak winter conditions on the Heuberg, or, as one source claims, because of a fire at the Schmidding-Werke factory had delayed delivery of booster rockets at Waldsee) the trials did not resume until in mid-February 1945 with the successive launches of M31 and M32. The general characteristics of these machines were the same as M17 and the trials were still without the Walter engine. The two launches demonstrated that the method of salvaging the Natter by parachute was still

Wolfgang Offik (right), manager of development and construction matters for the Natter project, and his deputy, H. Michelson (left), are holding a piece of the tailplane with the 'finder-reward' message on it. (M. Hensel)

very unreliable as M31 and M32 both crashed. The tests showed however that the aircraft could be launched into vertical stable flight from a guiding ramp of only ten metres.

The first Walter engines were delivered at the Bachem-Werk at the end of December and prototype M22, completed early in February, was the first to be built with the complete propulsion system. It was assembled and delivered to Heuberg and by February 25 was ready for launching.

The Walter engine was started and built up to full thrust. The booster rockets were then ignited and M22 roared up the ramp. The motor functioned perfectly despite the unusual vertical launch position and strong acceleration. At the given altitude, the nose separated and the dummy pilot was released and floated down under its parachute. The two salvage parachutes opened and took the fuselage and the Walter motor safely to earth. Unfortunately, after such a perfect test, some remaining propellants ignited and the machine burnt out on the ground. Nevertheless, at last there was the proof that the radical flight profile was feasible.

In January 1945 an 'Arbeitstab Dornberger' under Generalmajor Walter Dornberger was established at the Ministry of Munitions, in charge of all V-weapons and anti-aircraft development except for actual AA guns. Meeting at the end of the month to analyse the numerous guided missile projects then being developed in parallel, the 'Arbeitstab' only retained

Walter motors were finally delivered at Bachem-Werk and the first machine equipped with the rocket motor, prototype M22, was completed at the beginning of February. As no colour photos exist, the surviving Natter at the Smithsonian Institution may be a basis for proposing that the camouflage basecoat was RLM 76 light-blue and RLM 71 dark-green over it; underside was seemingly painted RLM 23 red, as were the nose cap, the bands at the separation points and the numbers on wings. (M. Hensel)

Final preparation of M22: technicians are installing the steel cover to protect the booster ignition cables from the exhaust from the Walter engine. According to Horst Lommel, author of Der erste bemannte Raketenstart der Welt, Geheimaktion Natter, the man at left with civilian hat is Willy Fiedler. Weighing 2,050 kilograms with the dummy pilot installed in the cockpit, prototype M22 was to be the heaviest Natter launched up to that time. The test flight of M22 on February 25 was a major step forward in the Natter programme, with the Walter engine and the four boosters all functioning perfectly, as well as the salvage parachutes. (M. Hensel)

three anti-aircraft rockets: the Wasserfall, the Schmetterling, and the X-4, a small wire-guided air-to-air missile. There was no mention of the Natter. However, pointing that the Natter required only a short development before being put into action, SS-Gruppenführer Kammler ordered on February 6 that the Natter project be continued. (SS-Gruppenführer Dr.Ing. Hans Kammler was Himmler's plenipotentiary for all matters relating to the A 4 rockets from the summer of 1944.)

FIRST MANNED FLIGHT

EARLY IN 1945, co-founder Dipl.Ing. Willy Fiedler joined the Bachem-Werk to assume the position of Chief Designer. Previously working for Fieseler, he was the leader of the industrial testing of the V1 and was responsible for the construction and testing of the projected manned version of the V1. He was known to be able to grasp and analyse the most difficult technical problem and gifted to improve any design up to perfection.

In January the SS, the most senior client of the Natter project, ordered a manned launch to be carried out by the end of February. Bachem protested that further tests were required but, as he later wrote, 'at this stage the customer provided a pilot, Gefreiter Lothar Siebert [sic], a dashing young officer who had volunteered for the mission'.

On March 1, the camouflage-painted prototype M23 was readied for launch with its tanks cautiously topped up with C-Stoff and T-Stoff. Finally weighing in at 2,540

Lothar Sieber was born on April 7, 1922 in Dresden. When he was 18, he volunteered to join the Luftwaffe and signed up in January 1940. One year later he graduated from the Instrument Flying School at Brandis and was promoted Leutnant in April 1941. Dissatisfied with his successive posts in Germany, he asked to be posted to a front-line unit and in November 1942 he was sent to Kampfgeschwader 51, a bomber unit on the Eastern Front. Early in 1943, after a farewell celebration with comrades, he failed to turn up for guard duty and was subsequently charged with being drunk on duty. On February 11 a court-martial at Luftgau-Kommando Moskau, Luftflotte 2, demoted him to the rank of Flieger (Private), and sentenced him to four months' imprisonment but, following an appeal, the sentence was reduced to six weeks to be served at Bobruisk.
(S. Gienschen)

In August 1943 Sieber was posted to the 2./Versuchsverband Ob.d.L., a unit tasked with operating undercover flights to insert agents into the Soviet Union. A posting followed in January 1944 to Kommando Süd, a unit flying clandestine missions from Greece to Libya. He was soon involved in Operation 'Etappenhase' which had the purpose of establishing clandestine airfields from which commandos would operate against Allied air bases in North Africa. On March 3 Sieber took off from Kalamaki in a Savoia Marchetti SM 75 to deliver fuel and supplies to the clandestine 'Lager I' at El Makarem, 90 kilometres south of Sirte, in Libya. However, he failed to find the objective and, after nearly seven hours of flight, he had to land in the desert for the night. There were three humps nearby so they called the place 'Drei Gräber' (three graves). Next morning, they took off and soon found the airfield where the supplies were unloaded, returning to Kalamaki on March 5. Sieber flew back to 'Lager I' on March 10 with the SM 75 in company with a Heinkel He 111. He took off with the He 111 in company with a Messerschmitt Bf 108 on the 12th to go to 'Lager II', where they landed. He then flew to 'Lager III' on the 13th. However, the following day British troops surprised the four-man party they had left at 'Lager I' – Obergefreiter Karl Heinz Fleckner, Obergefreiter Wilhelm Witte, Feldwebel Hubert Schulz and Unteroffizier Martin Ollhof) – capturing them and destroying the SM 75. When the He 111 and the Bf 108 flew back to 'Lager I' in the afternoon, there was no one left but the burnt-out wreck of the SM 75. They topped up the tank of the Heinkel He 111 from a concealed fuel reserve, destroyed the Bf 108 and flew back safely to Kalamaki on the 15th.

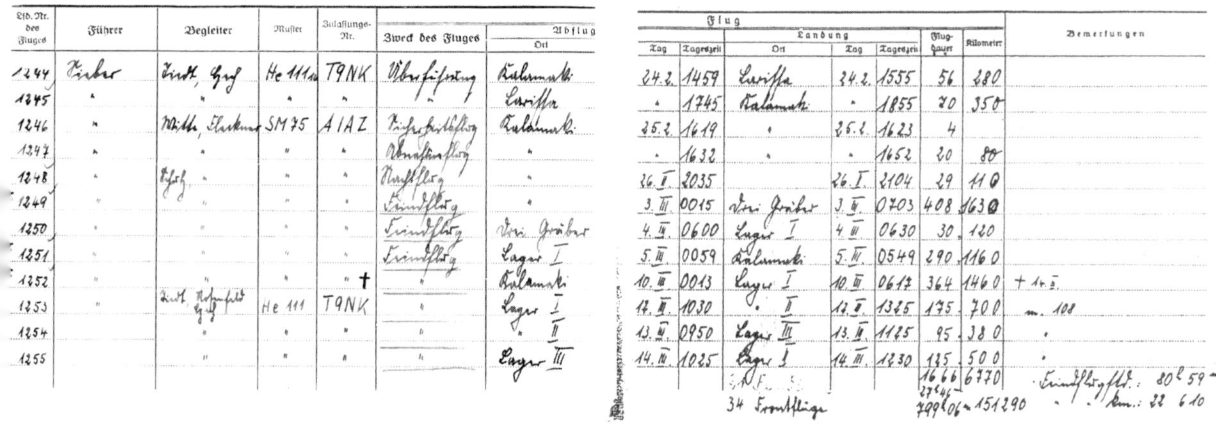

Lfd. Nr. des Fluges	Führer	Begleiter	Muster	Zulassungs-Nr.	Zweck des Fluges	Abflug Ort	Abflug Tag	Abflug Tageszeit	Landung Ort	Landung Tag	Landung Tageszeit	Flugdauer	Kilometer	Bemerkungen
1244	Sieber	[illegible], Graf	He 111	T9NK	Überführung	Kalamaki	24.2.	1459	Larissa	24.2.	1555	56	280	
1245	"	"	"	"	"	Larissa	"	1745	Kalamaki	"	1855	70	350	
1246	"	Witte, Fleckner	SM 75	A1AZ	[illegible]	Kalamaki	25.2.	1619	"	25.2.	1623	4		
1247	"	"	"	"	[illegible]	"	"	1632	"	"	1652	20	80	
1248	"	Schulz, "	"	"	Nachtflug	"	26.II.	2035		26.II.	2104	29	110	
1249	"	"	"	"	Feindflug	"	3.III.	0015	Drei Gräber	3.III.	0703	408	1630	
1250	"	"	"	"	Feindflug	Drei Gräber	4.III.	0600	Lager I	4.III.	0630	30	120	
1251	"	"	"	"	Feindflug	Lager I	5.III.	0059	Kalamaki	5.III.	0549	290	1160	
1252	"	"	"	" †	"	Kalamaki	10.III.	0013	Lager I	10.III.	0617	364	1460	+ 14.III.
1253	"	[illegible], [illegible], Graf	He 111	T9NK	"	Lager I	12.III.	1030	" II	12.III.	1325	175	700	m. 108
1254		"	"	"	"	" II	13.III.	0950	Lager III	13.III.	1125	95	380	"
1255		"	"	"	"	Lager III	14.III.	1025	Lager I	14.III.	1230	125	500	"
									34 Frontflüge			1666 / 27h46 / 799h06 = 151290	6770	Feindflugzeit: 80h 59m; " km.: 22 610

These two pages from Lothar Sieber's log book show his eventful flights in February and March for which he was awarded the Iron Cross Second Class and promoted Gefreiter. The first two flights were from Kalamaki to Larissa and back with the He 111 on February 24; then two local flights from Kalamaki with the SM 75 on the 25th and one on the 26th; he then flew to 'Drei Gräber' on March 3; he flew to 'Lager I' on the 5th, and returned to Kalamaki on the 6th; the last four lines show the flight to 'Lager I' on the 10th, to 'Lager II' on the 12th, to 'Lager III' on the 13th and the return flight to 'Lager I' on the 14th. (S. Gienschen)

Operation 'Etappenhase' ended in a failure by May and Lothar then returned to the Eastern front. Throughout the summer of 1944, he flew with Kampfgeschwader 200, mainly involving the delivery of agents behind the Russian front. His log-book for this period is filled with several mysterious entries including 'Heidelbeere', 'Melone', 'Reiher', 'Maulbeere', 'Tollkirsche', 'Herbstlaub', 'Pelikan', 'Traube', 'Feldspat', 'Kupfer', 'Schwefel'. In August he landed an Arado Ar 232 in the Ukraine to rescue 23 soldiers under enemy fire. (S. Gienschen)

On November 25, 1944, Lothar, by then a Gefreiter, became engaged to Gertrud Nauditt, who was a Luftwaffen-Helferin (civil employee of the Luftwaffe), the wedding being announced for July 5, 1945. On December 22 Sieber joined the Bachem-Werk as test pilot, subsequently observing a few of the test launches at Heuberg. (S. Gienschen)

kilos with propellants and pilot, it was to be the heaviest Natter to be yet launched.

At 11 a.m. all was ready and the signal was given. At first the launch looked perfect but when the Natter had risen to 100 metres it rolled into the inverted position and continued to climb at an inclination about 60 degrees from the horizontal. The cockpit hood then flew off and some 15 seconds after leaving the ramp, the engine cut out. M23 continued to climb but at about 1,500 metres was lost to sight in the cloud.

The sound of an explosion was soon heard in the distance. All eyes searched in vain for Sieber on his parachute but after an hour of waiting for news that he had landed safely, the launch team started a search. After another hour, the crash site was found in a forest south of Nusplingen, 6.3

March 1, 1945, dawned misty when prototype M23 was being prepared for launch at Heuberg. Note how the pilot head rest was attached to the cockpit canopy. (M. Hensel)

The launch scenario has been recreated in a full-size replica at the Truppenübungsplatz Heuberg museum. (JP. Pallud)

Here it was. In February 1999, the Bundeswehr carried out a detailed search to look for the foundation of the launch tower. This was featured in Der vergessene Raketenflug - Chronik einer tragischen Pioniertat, a documentary on the Natter project by the German TV channel Südwestrundfunk. (JP. Pallud)

Test pilot Sieber discussed final take-off instructions with Erich Bachem, the inventor, just prior to the launch of M23.

kilometres away. A small crater five metres deep marked the spot, the Natter having been totally destroyed. Of the pilot, only a piece of skull and half of his left arm and leg could be found.

Lothar Sieber was buried with full military honours in the cemetery at Stetten am kalten Markt on March 3 in the presence of Generalmajor Helmut Besch, the commander of Truppenübungsplatz Heuberg.

Reconstruction of the flight, which had lasted only 55 seconds, calculated that the Natter had reached a speed of around 800 km/h. A witness of the

Then the ground crew began to fill the tanks with propellants. C-Stoff truck appears on the right. According to Horst Lommel, the man in the hat at the top of the gangway is Erich Bachem. (USNA)

After climbing the gang-plank resting on the leading edge of the wing of M23, Sieber prepares to swing himself into the cockpit. No colour photo of M23 has come to light but black and white photographs indicate that it was finished in camouflage with a single colour underneath. As far as the replica M23 displayed in the Heuberg museum is concerned, the camouflage was based on that of the surviving Natter in the Smithsonian Institution with RLM 76 light-blue as base coat and RLM 71 dark-green over it; the numbers on the wings, the bands at the separation points and the nose cap were painted red, as was the underside. (USNA)

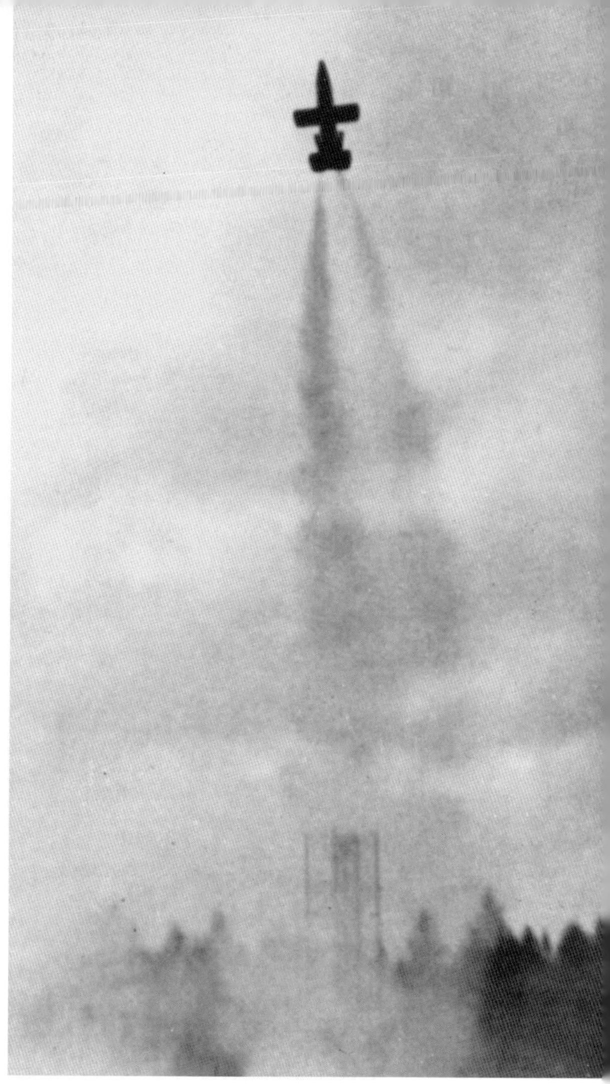

Karl Mielenhausen, a member of the ground crew at the Heuberg, later recalled this fateful morning of March 1, 1945: 'Towards 1100 hours M23 is ready for take-off. Low stratus cloud lies over the Ochsenkopf. The last mechanics leave the take-off area. The take-off signal is given with signal rockets.' Prototype M23 rocketed up the launch tower and continued to climb until it was lost to sight in the cloud base. Having reached an altitude of 1,500 metres, prototype M23 crashed after a flight of 55 seconds at the edge of a forest south of Nusplingen, about seven kilometres as a crow could fly from the launch point at the Heuberg. Lothar Sieber, the first man ever to take off vertically from the ground under rocket power, was killed. (USNA)

In 1980 the Bundeswehr erected a memorial to commemorate the first vertical take-off manned rocket flight in history and it was placed alongside the surviving concrete pad of the experimental pole launcher. Unfortunately, as the memorial lies well inside the military camp, it is not accessible to the public. (JP. Pallud)

flight, Dipl.Ing. Hans Zacher, recalled hearing the then unfamiliar double thunderclap, typical of a sonic boom during the dive, so Sieber might well have been the first human to break the sound barrier.

With so little wreckage to go on, and no film after the Natter entered the clouds, the analysis of the accident failed to explain precisely what had happened. The report by Bachem-Werk argued that under the 3g acceleration, Sieber might have involuntary pulled back the control column, causing the aircraft to tilt onto its back. The report suggested that in this upside-down position, his weight would have pushed – as far as his safety harness allowed – heavily against the cockpit canopy causing the latch to give way. (Another explanation not mentioned in the wartime reports, but which appears to emanate from the interrogation of Dipl.Ing. Bethbeder, is that the canopy might have opened simply because it had not been properly latched. As the pilot's padded headrest was attached to the underside of the canopy, the blow of it flying off might well have been sufficient to daze Sieber, knock him out, or even to have broken his neck.)

The Bachem-Werk report includes the suggestion that the stopping of

Truppenübungsplatz Heuberg, the launch site as it appears today. (Google)

The cockpit canopy that broke away from Prototype M23 a few seconds after launch was recovered after the flight. It was found that the starboard side was bent diagonally across the side window panel, damage caused when the canopy crashed to the ground. (M. Hensel)

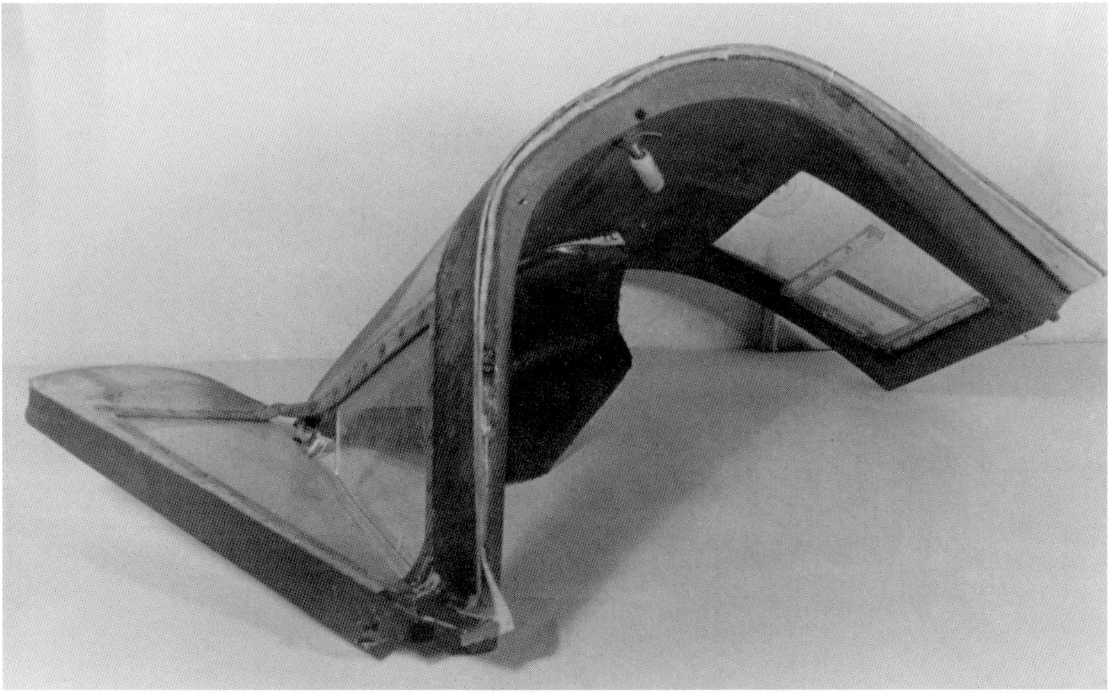

All that could be found of the pilot was his left hand with a piece of forearm and his left leg that had been ripped off below the knee. After examination of the five-metre-deep crater, a piece of skull was also recovered. Posthumously promoted to the rank of Oberleutnant, Lothar Sieber was buried with full military honours in the cemetery of Stetten am kalten Markt on March 3 in the presence of Generalmajor Helmut Besch, the commander of Truppenübungsplatz Heuberg. (JP. Pallud)

The grave of Lothar Sieber – No. 48 in the plot reserved for military burials – is now marked by a simple flat stone engraved with his name. (JP. Pallud)

the Walter motor 15 seconds after take-off could be due to Sieber turning it off himself, implying the possibility that he had only been slightly dazed by the hood breaking off. However, the Walter motor was notorious for shutting down unpredictably because of bubbles in the propellant feed lines. The report did not exclude the possibility that Sieber, dazed but conscious, may have lost orientation in the cloud and, trying to level out, instead went into the fatal vertical dive. Nevertheless, the Bachem-Werk weekly report makes it clear that, in spite of the tragic outcome, much had been learnt from the trial.

However, it would appear that the Bachem team's investigation was not quite as thorough as it should have been as excavations of the impact site in 1998-99 found the remains of one of the Schmidding booster rockets that had failed to release. This could have been an important factor in de-stabilising the Natter.

Manfred Hensel, the chairman of the Stetten Historical Association (Geschichts- und Museumsverein Stetten am kalten Markt / Heuberg e.V.), addresses those gathered for the remembrance ceremony held in March 2000. (M. Hensel)

In July 1998, Manfred Hensel obtained permission from the owner of the land to allow a one-day dig on the crash site of M23 in the Nusplingen forest. At a depth of over three metres, the largest piece of wreckage that was recovered was the steel ballast that was once mounted in the nose of the prototype to make up for the weight of the rocket armament. Another discovery was the remains of a Schmidding booster rocket. All four rockets should have been released when burnt out ten seconds after take-off and it was only now that it was obvious that one of them had failed to separate during the test flight, a fact that escaped the investigators' attention in 1945. The harness from Lothar Sieber's parachute was also unearthed, and sizeable pieces from the Walter rocket motor, some sections of fuel tubing, and many other smaller fragments. (M. Hensel)

The Natter crash site as the author saw it during his visit in 2010. In its final dive to oblivion – the speed at the end of the flight over 1,000 km/h – the machine descended diagonally across these trees, from top right down to left. Some trees were smashed but those only slightly damaged still show traces of the impact high on their trunks. (JP. Pallud)

Today new trees and bushes have totally reclaimed the area and nothing remains to be seen of the crater. (JP. Pallud)

OPERATION 'KROKUS' – TOWARDS OPERATIONAL LAUNCHES

THOUGH THERE CLEARLY were some vacillations in higher quarters as to the continuation of the project, Bachem pressed on with more trials, especially to test automatic guidance and a simpler launch mounting.

Prototype M24 was launched on March 15. This time the flight went according to plan: the booster rockets were released, the nose separated, and the parachute opened, bringing the Walter engine safely to the ground.

Fitted with a new three-axis auto-pilot, the M14 was the first Natter fitted with automatic guidance. Launched on March 16 powered only by the four booster rockets, it went into a spiral ascent just after having cleared the tower. The safety parachute opened but the machine was completely destroyed when it hit the ground. Prototype M34 was launched on March 21 for the same trial flight, and the outcome were more or less the same. It was only then that it was realised that the spiral flight pattern was because the autopilot had not been properly programmed to fit the specific aerodynamic characteristics of the interceptor.

Prototype M13 was launched on March 30, seemingly to test the rearward attachment of the booster rockets for the B-type series. Only three of the four boosters ignited but the take-off was still perfectly stable. One final launch from the tower at Heuberg was made on April 10 of M25 which had the Walter motor in place. The prototype took off but the four boosters failed just after the tower had been cleared. As the Walter engine did not have sufficient thrust on its own to lift the craft, it flew inverted up to 200 metres before crashing not far from the launch pad.

The steel launch tower was too costly to be produced in large numbers for operational launches so a simple wooden pole arrangement was devised. The trials had already proved that a shorter tower was sufficient to obtain a stable take-off so the new launcher was to extend 16 metres above the ground, the wooden pole having been sunk two metres into a concrete foundation. Providing for the necessary space at the bottom to mount the Natter with its rocket engine exhaust nozzle about 1.5 metres above the concrete foundation, this would give a launch ramp 12.5 metres long.

In the steel launch tower used for the first flight tests, the aircraft wing tips were inserted in lateral guideways, but the pole launcher provided a central channel for a short pylon mid-fuselage and another on the tail fin, and simple runners under each wing-tip.

A launch 'tower' of this new type was erected in March in the vicinity of the steel one at Heuberg and the first four machines of the production series – the A1 operational type – were readied for testing.

Prototype M52 was successfully launched from the pole launcher at the beginning of April, the flight faultless with both the aircraft and the dummy pilot floating to the ground on their parachutes. Although records are not complete as to the final sequence of trial launches, another one, equally successful, is believed to have been conducted some days later. In 1952, Erich Bachem recalled that seven launches were carried out after Sieber's death: four via the steel tower and three from the pole. However, the records state that there were five launches from the steel tower which would equate to just two from the pole.

In late February it was decided to establish an operational launch site near Stuttgart and Major Edmund Gartenfeld, former commander of Kampfgeschwader 200, was tasked with selecting eight experienced pilots

Prototype M52 – an experimental version of the A1 operational type – was prepared for take-off at Heuberg. (M. Hensel)

The prototype of the simpler pole launcher was erected in the vicinity of the steel launch tower. Differing from the launch tower where the aircraft wing tips were inserted in lateral guideways, the pole launcher provided runners under each wing-tip and a central channel for the shoe mounted mid-fuselage on a short pylon and the one on the tail fin. (M. Hensel)

Engineers discuss the launch of Prototype M52 (in the pole launcher in background). Manfred Hensel identified test pilot Hans Zübert (sat in right forefront) among them. (M. Hensel)

Photos were taken for future instruction of the operational crews. The fuelling of the Natter with C-Stoff was a delicate and dangerous operation, the ground crew man operating was dressed in a protective gown and he had to manoeuvre the propellant line with gloved hands. (M. Hensel)

for this operation. Code-named 'Krokus', the deadline for going into action was March 20.

Although the weapon itself has been developed, until now no real work has been done concerning target location. Consequently Bachem-Werk now put in an urgent request for Flak specialists and at the end of February a construction team from the Organisation Todt was ready to build the concrete foundations for the pole launchers. The plan was for three poles to be set up at the points of an equilateral triangle which had sides 100 metres long. The range-finding apparatus and the Malsi-Gerät, a radio beam

Prototype M52 was painted mainly in RLM 02 grey-green with the nose cap and booster rockets in RLM 23 red. To help in identifying the machine's orientation during flight, large geometric symbols were painted in yellow on the top and bottom of each wing: a cross on the upper right; a circle on the upper left; a slash on lower right, and a triangle on lower left. (M. Hensel)

projector that would guide the three Natter interceptors towards their targets, was to be located in the centre. The fabrication of ten pole launchers was ordered from the engineering firm at Esslingen which had already built the steel tower at Heuberg and by the end of February a team from the Organisation Todt was building the concrete foundation for three pole launchers.

Back in 1955, author Karl Bartz wrote in his book *Als der Himmel brannte* that the operational launch site for the first Natters was located at Kirchheim

Prototype M52 was successfully launched from the pole launcher at the beginning of April, the flight faultless with both the aircraft and the dummy pilot floating to the ground on their parachutes. (M. Hensel)

At Heuberg, the concrete foundation (in the shadow just to the right of the memorial stone in the background) where the experimental pole launcher was erected has survived making it possible to take an accurate comparison. (JP. Pallud)

unter Teck, 20 kilometres south-east of Stuttgart, and he wrote that 'until recently' remnants of the burnt-out machines were lying 'scattered around at the glider airfield'. This account was accepted as gospel even though it was not backed up by documents or photographs. Then in 1998, following the publication of Horst Lommel's book Der erste bemannte Raketenstart der Welt, Geheimaktion Natter, Lommel was contacted by Willy Reichert who told him about some concrete slabs he remembered seeing in 1945 in a wood near the town of Holzmaden, a few kilometres south-east of Kirchheim. They investigated the woods together in 1999 and soon discovered three foundations which were arranged in the form of an equilateral triangle, the distance between the launch pads being approximately 50 metres. What had been relayed to Karl Bartz about Natter's launch ramps at the local airfield at Kirchheim unter Teck had simply been remnants of structures used to launch gliders.

At Heuberg, even the end of the wooden pole remains set in the concrete. (JP. Pallud)

The location of the long-forgotten launch site was achieved in 1999 following the publication of Horst Lommel's book *Der erste bemannte Raketenstart der Welt, Geheimaktion Natter*. Researching for the book, Horst Lommel met Willy Reichert who told him that he remembered having seen some strange concrete slabs in a forest south-west of Holzmaden in the autumn of 1945 when he was working as a trainee surveyor. Reichert took Lommel to the wood and there they discovered the three foundations. This was the small Hasenholz wood as it was in 2010 when the author visited the site. (Google)

At Kirchheim unter Teck, in the Hasenholz wood, this is Pad 'A' as pictured by the author in 2010. Note that the letters 'A', 'B' and 'C' are merely our own references. (JP. Pallud)

The three blocks were arranged in the form of an approximate equilateral triangle, the distance between them was between about 100 metres (between Pads 'A' and 'B', and between 'A' and 'C') and about 120 metres (between 'B' and 'C'). In the centre of each pad was a square recess, about 50 centimetres deep, for the wooden pole. The end of a pipe, now cut off at ground level, would have carried the firing cables. This was Pad 'B' which formed the eastern tip of the triangle. (JP. Pallud)

This is Pad 'C' which formed the southern tip of the triangle. (JP. Pallud)

Sadly, Pads 'A' and 'B' were removed in 2015 during construction work on the new Wendlingen – Ulm high-speed railway line that runs parallelly to the A8/E52 motorway. Work began in 2012 and the line was officially opened in December 2022. In February 2015, ufo709 took this picture of pad 'A' being cleared from the ground to be removed. (ufo709)

In 2017, Martin Merinsky photographed (above) what was left of Pads 'A' and 'B' stored in an open place. He also took this photo (left) of Pad 'C', the only surviving pad on the site, now marked with an information board.
(Martin Merinsky)

We can see the remains of Pads 'A' and 'B' stored in the centre of this aerial view from 2018. (Google)

Fortunately, while Pad 'A' appears to have been destroyed, the remains of Pad 'B' were taken away for preservation by the Militärgeschichtlichen Sammlung (Museum) at Stetten am kalten Markt. The pieces were temporarily stored in the open when we finalised this book in late 2023, but they will soon be included in the museum. The pieces on the left in this picture were parts of the circular platform at ground level, the largest part on the right being the central core of the launch pad, which was two metres deep; the lower part of this core has not been removed, and it is still buried in the ground at Holzmaden. (Militärgeschichtlichen Sammlung Stetten am kalten Markt)

FINAL PRODUCTION AND CAPTURE

AT THE BEGINNING of March, a Bachem-Werk schedule listed 15 A1 operational interceptors with production deadlines from March 23 to April 20. These were to be fully-armed with the Föhn rocket canister in the nose and simple gunsight. Erich Bachem later said that 14 had been completed or almost completed before the war ended and that all in all, 36 Natters were built.

Meanwhile a contract had been signed with a sub-contractor, Wolf

Combat Command 4 of the 5ème Division Blindée entered Waldsee late on the afternoon of April 24. In this picture, Sherman tanks of the 1er Régiment de Cuirassiers sped eastwards across the town. (ECPAD)

The historic city of Waldsee is situated on an isthmus between two lakes, the Stadtsee on the east and the smaller Schlosssee on the west. This is the Hauptstrasse looking west from the Heilig Geist Spital Church. (JP. Pallud)

The Shermans were followed by half-tracks of the Régiment de Marche de la Légion Etrangère. After the war the city of Waldsee developed its tradition of spas and changed its name into Bad Waldsee, 'Bad' meaning 'spa' in German. (ECPAD)

Dipl.Ing. Hans Zacher is claimed to have sunk some 10 to 15 Walter propulsion units in the Stadtsee lake before the French force reached Waldsee. However, when an underwater search was conducted in the late 1960s under the auspices of the Deutsches Museum, nothing was found. Possibly they were recovered just after the war by waste metal dealers and scrapped. It is possible that a French photographer heard about this story and then took this photograph of the lake in April 1945. (ECPAD)

The rowing clubhouse still stands on the eastern side of the Stadtsee. (JP. Pallud)

Hirth Flugzeugbau at Nabern unter Teck, three kilometres south-east of Kirchheim unter Teck, for the pilot production of ten interceptors, and three partially completed Natters (said to be 50 to 75 per cent complete) were discovered at this factory at the end of the war.

Plans were also under way for production at another factory in Thuringia under the cover name of 'Waldhausen'. Horst Lommel identified this factory as being the Kühn & Lefler Büromöbelfabrik, a furniture factory, at Gräfenroda, 30 kilometres south-west of Erfurt. A launch tower was to be built at Trüppenübungsplatz Ohrdruf, ten kilometres to the north-west, to test the output.

Some confusion surrounds the last days at Waldsee. By mid-April, it seems that SS-Obergruppenführer Kammler issued an order to evacuate to Austria. A number of Natters were loaded onto lorries and trailers with some equipment and the convoy set off eastwards, travelling only at night because of the air threat. Bachem was not there when the convoy left, but Bethbeder left along with it. The convoy finally reached Bad Wörishofen, 80 kilometres west of Munich, where they remained for some days, apparently waiting to meet up with transport bringing a single Natter from the Wolf Hirth factory.

With French tanks approaching Waldsee, four Natters were taken to a meadow where they were burnt. (According to one story, 15 of the Walter propulsion units were dropped into the lake but a search conducted in the 1960s by the Deutsches Museum failed to find any of them.) On the afternoon of April 24, tanks of the 5ème Division Blindée entered Waldsee.

Remarkably, it is precisely at this same place that 'Eriba' took photos for the commercial brochure presenting their camping caravan Puck Luxus. After leaving Germany to settle in Argentina in 1947, Erich Bachem returned in 1952 and in September became technical director of the Ruhrthaler Maschinenfabrik Schwarz & Dyckerhoff KG in Mülheim. There he developed locomotives and various parts of mining machinery and served as technical director until his death in March 1960. From 1957, together with Erwin Hymer, he designed camping caravans under the brand name 'Eriba' that soon became a legendary name in the world of touring caravans

Another 'Eriba' advertising photo from the early 1960s, this one for the Faun Standard model. Eriba and Hymer worked together successfully until 1980 when the two companies merged. Today, Hymer is one of the most innovative and successful manufacturers of caravans and motorhomes in Europe, with a range of over 60 models.

On May 4, 1945, Company L of the 324th Regiment, US 44th Infantry Division, entered the Pitz valley, 60 kilometres west of Innsbruck, and within a day or so reached Sankt Leonhard. There they found four Natter interceptors lying abandoned by the side of the road. Though it is clear from the records that four of the A1 type were recovered from Sankt Leonhard, only three appear in the photographs and the cine film. (USNA)

In 2009 the author traced the exact spot where the four Natters were abandoned in 1945 – in a grassy meadow about half a kilometre before reaching the church of Sankt Leonhard. This part of Sankt Leonhard around Eggenstall No. 195 lies at an altitude of 1,360 metres.

The light attire worn by the GIs, and the fact that fields in background were neatly mowed, might indicate that the photos were taken later in May or early June. (USNA)

At Bad Wörishofen, one convoy started out with four Natters towards Austria. The second party with two Natters left Bad Wörishofen in the nick of time, just before lead elements of the US 10th Armored Division entered the town on April 27. With the Americans hard on their heels, the group managed to reach the Fern Pass where German forces were able to hold up the Americans for two days.

Meanwhile the first group had arrived at Sankt Leonhard in the Pitz valley, 60 kilometres west of Innsbruck, although the second convoy failed to meet up with them, having stopped near Ötztal, lower down in the Inn valley, some 30 kilometres away. The two Bachem interceptors at Ötztal were set on fire as the Americans approached on May 4 but the four at Sankt Leonhard were captured intact.

The Natter personnel from Schlatt and Sankt Leonhard were sent to a camp at Haiming in the Inn valley where they were interrogated by CIOS agents. (Established in the summer of 1944, the Combined Intelligence Objectives Sub-committee was tasked to plan and direct the exploitation of German technology, and it consisted of officers from the technical branches of British and American armed forces).

Captain R. W. Bratt of the US Army Air Corps and Flight Lieutenant J. R. Evans of the Royal Navy were assigned to investigate the Natter. They interrogated 13 Natter personnel, including Dipl.Ing. Bethbeder; Dipl.

Close-up of the nose section of the Natter in the centre of the three, showing the honeycomb canister for the 24 Föhn rockets. Although the outlines of the fields have changed, and many new houses erected, the lone chalet in the meadow remains. (USNA)

This photograph shows the side view of the Natter in the centre of the group, with a Schmidding booster rocket lying on a trailer. Also, the nose section from another Natter lies in the foreground. The calvary still remains standing at the side of the road. (USNA)

Ing. Richard Granzow, an engineer from the Walter-Werke detailed to the programme; Dipl.Ing. Gerhard Schaller, special representative of the Reichsführer-SS for the project; Unteroffizier Hans Zübert, the test pilot; and Dr. Heinrich Rieck of the DFS.

The Natter engineers believed that they could continue development of the BP-20 for the Allies, stating that demonstration flights could be carried out in two or three months providing they were given suitable facilities. However, reports by Bratt and Evans concluded that the BP-20 did not merit further investigation although they listed the documents and equipment to be shipped to the US and Great Britain.

The US Navy also fielded its own investigation team – Dr. Clark B. Millikan, Lieutenant C.L. Poor III, USNR, and Lieutenant A. Hyatt, USMCR – under the title of US Naval Technical Mission in Europe. In June 1945, the Naval team questioned six members of Bachem's workforce, all of them part of the original 13 Natter personnel already interrogated by Bratt and Evans. After having been held for some weeks at Camp Haiming, the group was moved to Jerzens, near Wenns, where they were under house arrest 'to be available for further interrogation should this be desired.'

The Navy Team reported:

> 'Bethbeder, who speaks and understands English fairly well, was interrogated for several hours and furnished most of the information on the project. He was very cooperative and apparently withheld no information on matters of interest. He impressed the team as an extremely ingenious and capable engineer. The project has so many unorthodox and unusual features that it is apt to appear unrealistic and crackpot. Bethbeder was able, however, to demonstrate that each of the unorthodox elements had been carefully worked out and engineered in accordance with a consistent' and very reasonable conception of the overall problem. (The data given by Bethbeder, 'from memory', were for the A1 operational Natter.)
>
> 'Zübert was the test pilot for the project and furnished considerable information on the flying characteristics of the glider version which he had flown.
>
> 'Granzow was Walter's representative on the project and was responsible for the functioning of the main rocket motor. He contributed relatively little to the team's knowledge.
>
> 'Gerhard Schaller was sent by RLM as a liaison man and inspector for development and production. It was stated that he might have been a Nazi party member, but that he had not been active in the party. His interrogation contributed little.
>
> 'Karl Schaller and Jonas were talked to only very briefly, since they apparently were workers who knew very little about the project.'

As explains the report, the team failed to catch Bachem despite their efforts:

> 'Bachem remained at Waldsee when Bethbeder's group evacuated

to the Inn Valley. He planned to bury a complete set of Natter drawings and technical data in the Waldsee area. It was agreed that he and Bethbeder would later attempt to establish contact by leaving messages at the latter's ski hut 'Einen Eckalpe' at Oberstaufen, 15 kilometres from Isny im Allgäu. The rendezvous has not yet been effected but Bethbeder believes that Bachem is probably in the neighbourhood of Isny, and is confident that he, Bethbeder, could find him.'

The Navy Team recommendations were:

'(a) The successful, vertical launching technique and the water-cooled jet vanes are of considerable technical interest. The pertinent information should be made available to persons concerned with interceptor aircraft, guided missiles, and rocket motors.

'(b) The flying characteristics and unusual armament warrant further study by airplane designers and 'military requirements' personnel.

'(c) The amount of Natter material turned over to Japan should be ascertained if possible. The use of Bethbeder to find Bachem for interrogation should be considered in this connection.

'(d) If it appears that Japan may be actively working on Natter, the difficult problem of counter-measures should be undertaken promptly.

'(e) Since Natter is essentially a defensive weapon, it is not likely that it should be directly useful to the US, in military or naval campaigns which are likely to occur in the near future. Certain of the special features Natter might, however, be usefully applied in the development of offensive weapons.'

In 1952 Erich Bachem claimed that one Natter had been captured by the Russians, namely the prototype that was sent from Waldsee to the new factory to be established in Thuringia. However, this area was first occupied by American forces before it was handed over to the Russians on July 1 because it was to become part of the Soviet occupation zone. It is therefore difficult to believe that the American intelligence teams would have missed a Natter when they expended so much effort during those weeks to remove German scientists from the clutches of the Soviets.

According to the CIOS report the following items were shipped to a collecting point near Stuttgart: two repairable BP-20s; two serviceable Walter rocket motors; four Schmidding booster rockets; four wings; two boxes of Föhn rockets; parachutes for the BP-20s, and a few boxes with miscellaneous instruments and fittings.

In trying to piece together the movements and final fate of the equipment despatched to the US, Brett Gooden (*Projeckt Natter, last of the Wonder Weapons*, 2006) named them Unit One and Unit Two. He traced Unit One – captured equipment number T2-1 – from Santa Monica in California in 1945, to Freeman Field, Indiana, in 1946, through to 1949 when the US Air Force transferred it to the Smithsonian Institution. The Natter, then in poor condition, was put into storage at Silver Hill in Suitland, Maryland (later known as the Paul E. Garber Facility). The Smithsonian National Air and Space Museum now plans to restore it and place it on display at the Steven F. Udvar-Hazy Center near Washington, DC.

The fate of 'Unit Two' is unknown but there is a possibility that it might have been flight-tested at Muroc Flight Test Base (now known as Edwards Air Force Base) in California in 1946 because a Natter is rumoured to have been launched from there in June under radio control although there appears to be no positive proof to confirm this.

A Signal Corps cameraman was also present in Sankt Leonhard and stills from his cine film provide us with detailed photographic coverage of the four Natters. The white-shirted civilian who appears to willingly demonstrate the technicalities of the Natter to US troops is believed to be Dr. Heinrich Rieck of the DFS. (USNA)

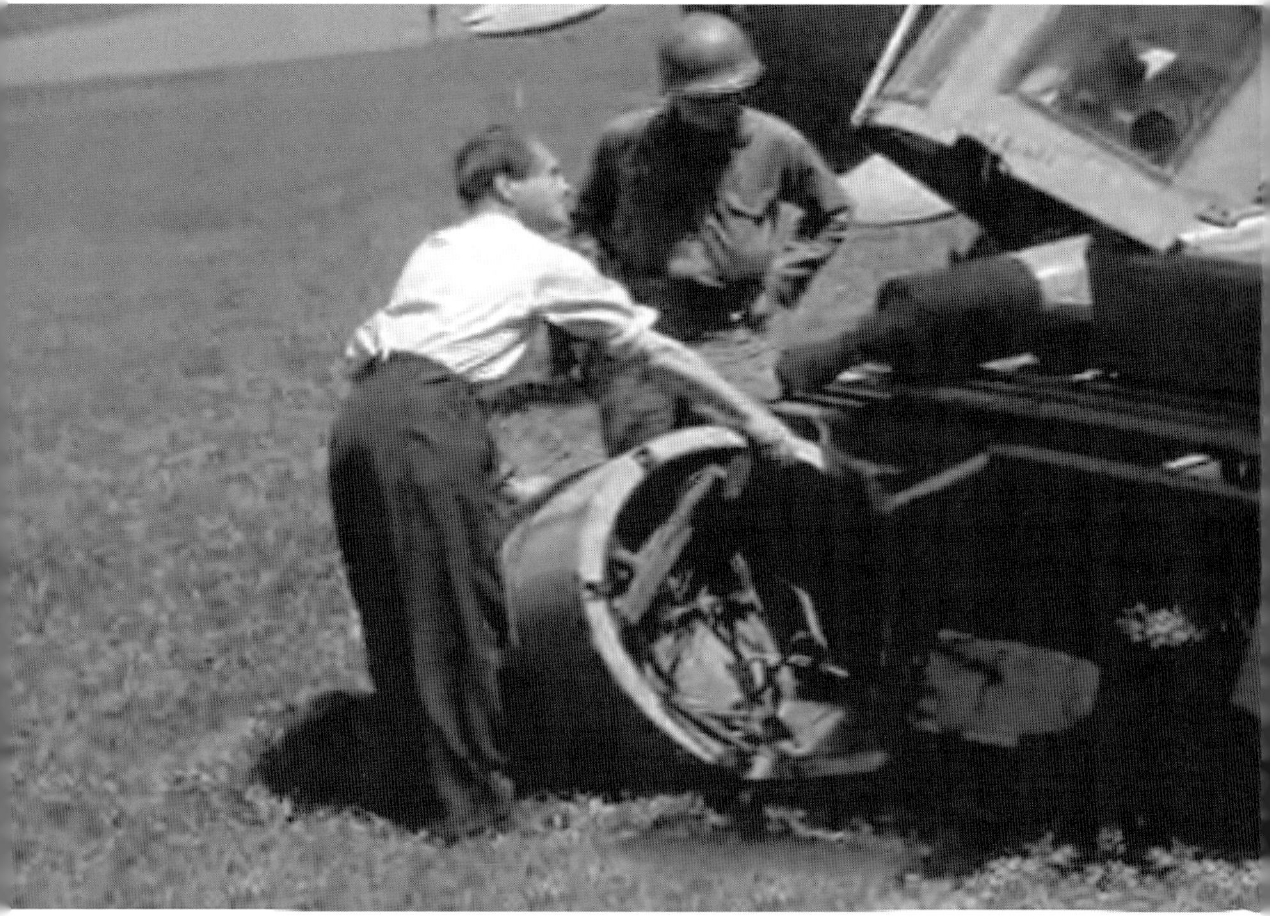

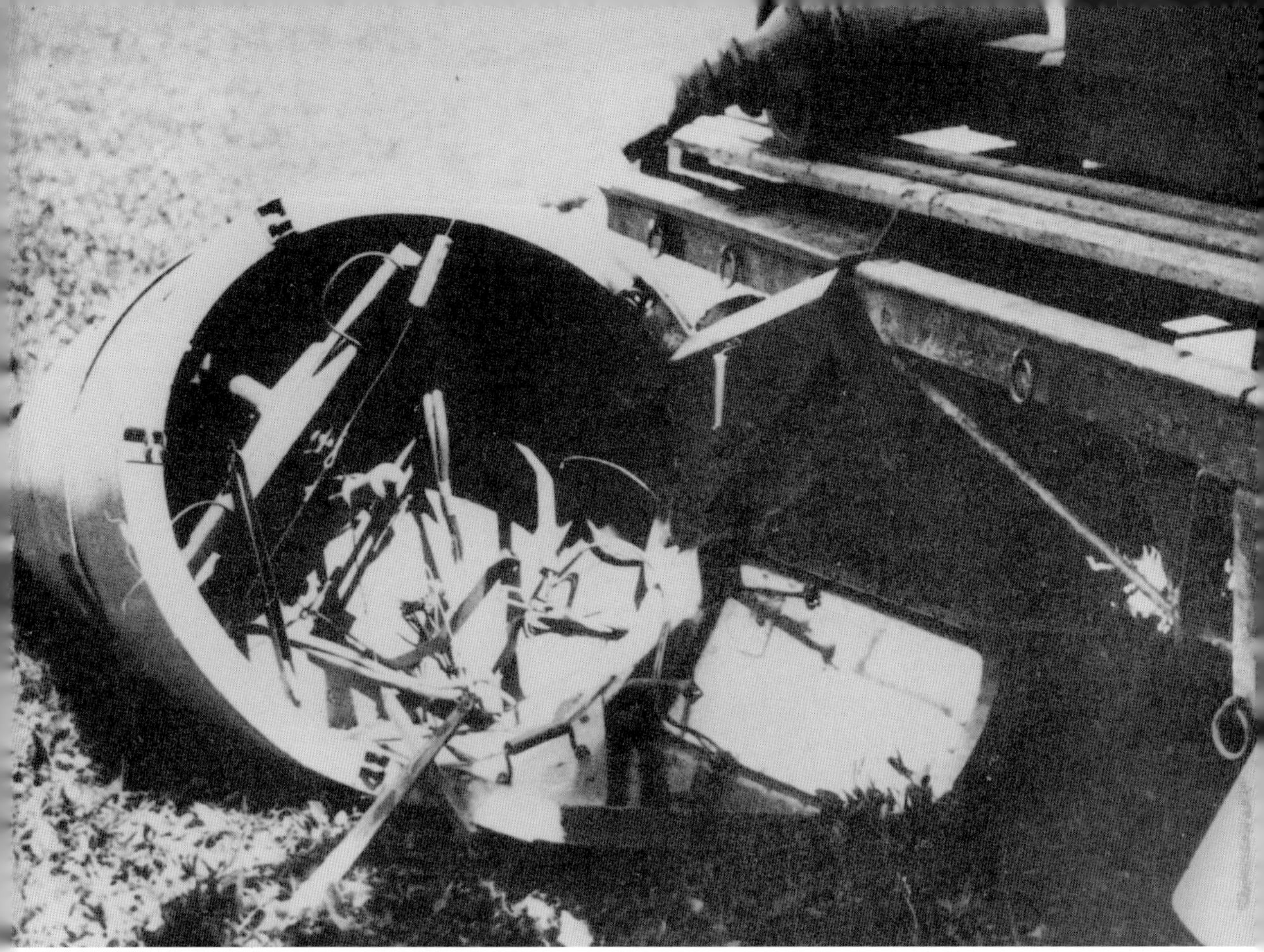

Heuberg museum

In 2006 the Stetten Historical Association loaned all the items found at the crash site for exhibition at the Militärgeschichtlichen Sammlung Stetten a.k.M., a museum that was then being set up at the Heuberg camp. The major exhibit is the replica of Natter M23 on its launch pad in full-size with models of Bachem and Sieber discussing final instructions just prior to the launch.

The Walter HWK 109-509 motor that equipped the Natter generated a maximum thrust of 1,700 kilos (16,7 kNewtons) but it could be throttled back to a minimum of 300 kilos.

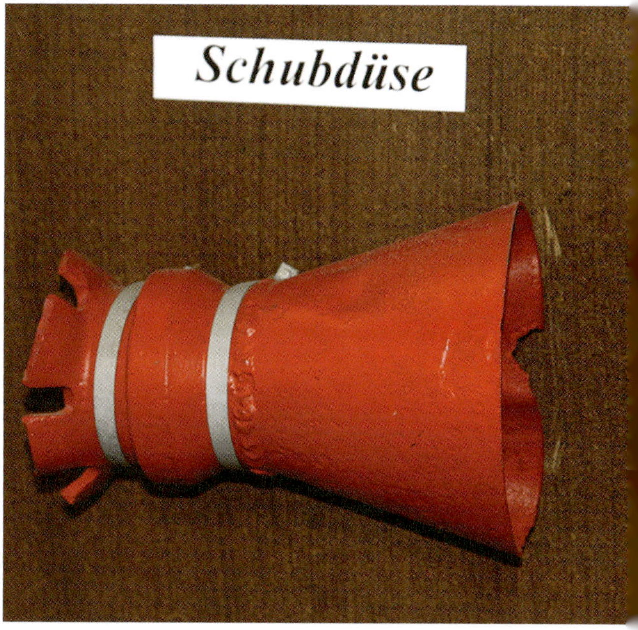

A replica of the tailplane of prototype M17 which was flight-tested in December 1944. The inscription in German reads: 'Finder report to Headquarters Troop Training Ground Heuberg. Telephone Stetten am kalten Markt 222. Reward.'

The exhaust nozzle of the Schmidding booster rocket also recovered at the crash site. It should have been released when burnt out ten seconds after take-off but it had failed to separate during the test flight.

The ballast from the nose of M23 recovered at the crash site. Weighing some 70 kilogrammes it was once mounted in the nose of the prototype to make up for the weight of the armament.

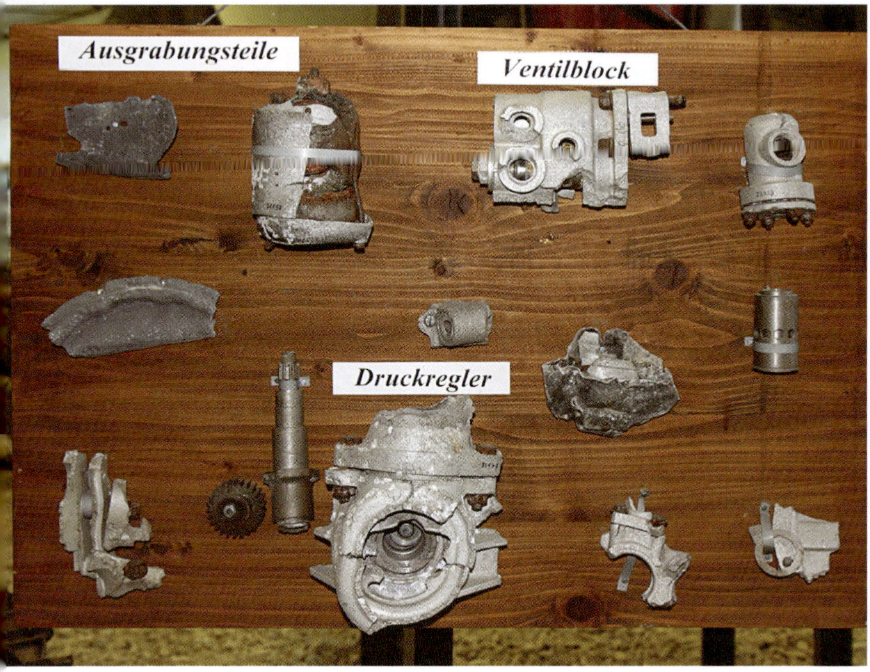

Parts of the Walter rocket motor recovered at the crash site are on display, including pressure regulator (Druckregler) and valve block (Ventilblock).

The most moving exhibit on display is the harness from Sieber's parachute which unfortunately is difficult to photograph well in its sealed glass cabinet.

A BMW R-71 with sidecar. The Militärgeschichtlichen Sammlung Stetten a.k.M. museum displays weapons, uniforms, and equipment covering over 100 years of military history at the training ground, from the days of the Imperial Army to the present-day Bundeswehr. Being located inside the closed military training area, the museum only opens to visitors every Sunday afternoon from 2 p.m. Guided tours can be arranged for groups or schools.

Other museums...

In his efforts to piece together the movement and fate of the captured Natters shipped to the United States, Australian author Brett Gooden named the two interceptors that he was able to track down as 'Unit One' and 'Unit Two'. This photo of 'Unit Two' on display at Freeman Field in September 1945 shows the triangular imperfection on the right nose, in front of the cockpit, that is one of the distinguishing features retained by Brett Gooden to identify 'Unit Two'.

Still strapped to the same pallet but now having the swastika on tail fin outlined Unit Two was photographed beside a four-engined Junkers Ju 290 at Wright Field Air Fair, Ohio, in January 1946. (USNA)

According to an original caption of 1949, this photo was taken in August 1946 at Wright Field and may well be the last known photo of Unit Two. Its final fate is unknown but one possibility is that it was flight-tested at Muroc Flight Test Base, California, later in 1946. (USNA)

'Unit One' was first traced at Freeman Field, Indiana, for analysis. where it was exhibited during Open Days in September 1945. It was moved to Orchard Place Airfield, Illinois, and then stored at an aircraft factory at Park Ridge. At some time during this period, its wings were cut through near the roots and hinged along the side to ease transportation. It was transferred to the National Air Museum in May 1949. 'Unit One' is currently in storage in Building 24 at the Paul E. Garber Facility in Suitland, Maryland, awaiting restoration. Eventually it is due to be displayed at the Steven F. Udvar-Hazy Center near Washingon, DC.

Another Natter is on display at the Deutsches Museum in Munich. It was rebuilt in 1971-72 from original parts recovered in 1967 from the Bachem factory. Painted in the yellow of prototype M17 which was flight-tested in December 1944, it shows a pattern of black stripes and the finders/reward inscription on the tail. (Anagoria)

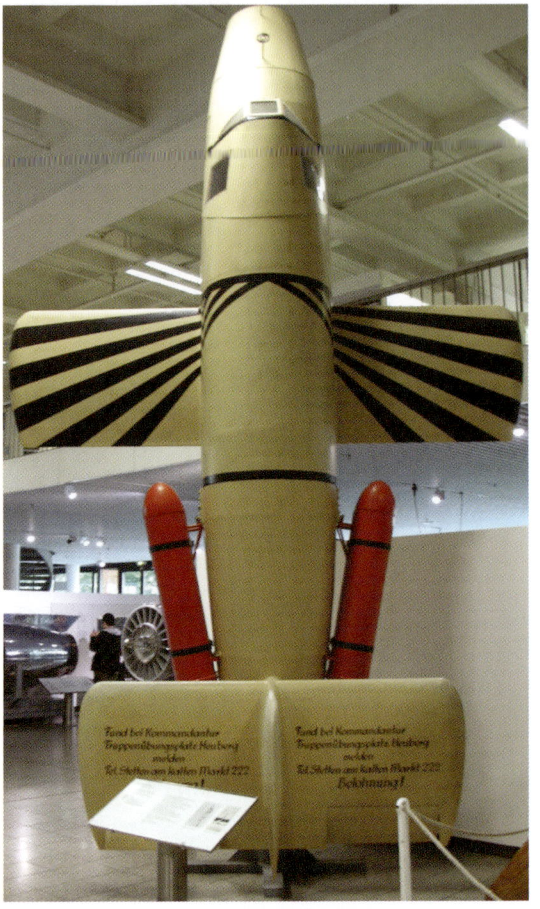

Natter replicas can be seen at the Planes of Fame Museum in Chino, California, (this photo) and at the Fantasy of Flight Museum near Orlando, Florida. (Senjūkaishi)